Fundamentals of Horticulture

— *Authors* —

Dr Mukesh Kumar

Assistant Professor (Horticulture)
CCS Haryana Agricultural University
College of Agriculture, Bawal (Rewari)

Dr Jeet Ram Sharma

Director of Research
CCS Haryana Agricultural University, Hisar

2025

Daya Publishing House®

A Division of

Astral International Pvt. Ltd.

New Delhi – 110 002

ISBN: 978-93-5919-552-0 (CCSHAU/PUB#23-104)

Publisher's Note:

Every possible effort has been made to ensure that the information contained in this book is accurate at the time of going to press, and the publisher and author cannot accept responsibility for any errors or omissions, however caused. No responsibility for loss or damage occasioned to any person acting, or refraining from action, as a result of the material in this publication can be accepted by the editor, the publisher or the author. The Publisher is not associated with any product or vendor mentioned in the book. The contents of this work are intended to further general scientific research, understanding and discussion only. Readers should consult with a specialist where appropriate.

Every effort has been made to trace the owners of copyright material used in this book, if any. The author and the publisher will be grateful for any omission brought to their notice for acknowledgement in the future editions of the book.

Published by : **Daya Publishing House®**
A Division of
Astral International Pvt. Ltd.
– ISO 9001:2015 Certified Company –
4736/23, Ansari Road, Darya Ganj
New Delhi-110 002
Ph. 011-43549197, 23278134
E-mail: info@astralint.com
Website: www.astralint.com

Prof. B.R. Kamboj
Vice Chancellor
CCS Haryana Agricultural University
Hisar – 125004, India

Foreword

Horticulture is very important component in agriculture sciences. It contributes about 30 per cent of agriculture GDP, 40 per cent of agriculture export on 14 per cent of total cultivable land. Its share is increasing at very fast rate but per capita availability of fruit in our country is still 207.9 g per day which is far below the recommended quantity of 230 g. Fruits and vegetables provide the carbohydrates, nutrients, vitamins *etc.*, and are aptly called as protective food and so indispensable part of human diet. It is also helpful to fulfil the greenery requirement to avoid global warming and provide shelter to insects, birds and animals.

Total horticultural production is excelling food grain production in India. Export of fresh fruits, vegetables and their processed products are in great demand in the foreign countries to earn foreign exchange *e.g.* fresh fruits and vegetables – apple, mango, grapes, mushroom, onion, potato *etc.*, flowers – cut flowers (roses), potted and foliage plants, plantation – cashewnut, tea, coffee, coconut, arecanut *etc.*

The book on Fundamentals of Horticulture provides thorough knowledge of agriculture/horticulture to the under graduate, and post graduate students, farmers/fruit growers. All the basic concepts for the production of fruits have been covered in this book. This book will also be useful for JRF/NET and other competitive examinations. Most of the techniques are explained with the help of pictorial form for better understanding of the readers. This book is compiled with latest comprehensive information which will be helpful not only for students but also for researchers, teachers, fruit growers and others working in the field of horticulture.

The book is enriched by including technical information related to production of horticultural crops such as importance and scope, classification, soil and climate, propagation methods and propagation structures, establishment of orchard, dormancy, juvenility, flowering, pollination, pollinator, fertilization, fruiting, unfruitfulness and fruit drop, training and pruning, methods of application of irrigation, manures, fertilizers and hormones *etc.*

It is hoped that this book will improve the knowledge and skills of the students/ farmers/fruit growers in the field of horticulture. I congratulate the authors and appreciate their efforts for preparing this book. This book is prepared in a simple language with pictorial representation for clarification of the concepts.

I appreciate the efforts made by the authors Dr Mukesh Kumar and Dr Jeet Ram Sharma to compile the information for the course Fundamentals of Horticulture as per syllabus approved by ICAR. I hope this book will be helpful to the students of UG and PG classes for preparation for different examinations, researcher, teachers, extension workers and fruit growers to empower their knowledge and contribute in the nation development and doubling farmers income.

Prof. B.R. Kamboj

Dr. S.K. Pahuja
Dean, College of Agriculture
CCS HAU, Hisar-125004
Haryana, India

Foreword

Horticulture is an important sector for diversification in agriculture. It is helpful to increase production, generation of employment and enhance export. The GDP contribution from horticulture is about 30 per cent of agriculture. India is second largest producer of fruits and vegetables in the world after China. Globally, India is at the first position in production of banana, mango, pomegranate, lime, lemon, papaya, coconut and cashew, and productivity of grape, banana, cassava and papaya. Fruits and vegetables are more resilient to change in weather conditions. This sector enables the large population to enjoy healthy and balanced diet. High production, its value, productivity and net profit from horticultural crops compared to food grains *e.g.* the yield of wheat/ground nut/*ragi* is about 3.0 - 4.0 MT/ha valued Rs. 25000/- to 35,000/- as compared to production of grapes/mango/banana is about 20-40 MT/ha valued Rs 1.5-2.5 lakh and the growth rate always remains higher compared to agriculture sector.

There is great scope for utilization of consumption of fruits within the country and to meet out export demand. The production and productivity can be enhanced by adoption of quality planting material, efficient utilization of land and water, knowledge about production technology, infrastructure, assured market rate. Fruit is one of the important inputs for the different fruit processing industries and other agro-industries –like canning and preservation of fresh fruits like aonla, bael, coir industries (coconut husk), pharmaceutical industry (aonla, bael, papaya, jamun) transporting and packaging industries (banana leaf and wood of other fruits) *etc.* Many horticultural produces and their byproducts are being used as raw materials for many industries, *e.g.* rubber, coir (coconut) and sago (tapioca) industries.

This book is enriched by including technical information related to production of horticultural crops such as importance and scope, classification, soil and climate, propagation methods and propagation structures, establishment of orchard, dormancy, juvenility, flowering, pollination, pollinator, fertilization, fruiting, unfruitfulness and fruit drop, training and pruning, methods of application of irrigation, manures, fertilizers and hormones *etc.*

I congratulate the authors Dr Mukesh Kumar and Dr Jeet Ram Sharma for preparing this comprehensive and informative course Fundamentals of Horticulture as per syllabus approved by ICAR. I hope this book will be helpful to the students of UG and PG classes for preparation for different examinations, researcher, teachers, extension workers and fruit growers to empower their knowledge and contribute in the nation development and doubling farmers income.

Dr. S.K. Pahuja

Dr. Naresh Kaushik

Principal
CCS HAU College of Agriculture
Bawal-123501, Rewari, India

Foreword

Horticulture plays an important role in nutritional security, employment generation and economic development, environment and now in international trade and overall growth of the country. Horticultural crops are grown in an area of 28.04 mha which contributes 347.18 m MT production, however food grain crops cover an area 129.8 m ha and contributes the production 310.7 m MT. The production of horticultural produce has increased 7-fold which ensured nutritional security and employment opportunities. Export growth of fresh fruits and vegetables in term of value is 14 per cent and of processed fruits and vegetables is 16.27 per cent.

Apart from yield, income and employment this sector also provides nutritional benefit. At the same time the horticultural crops are good source of vitamins and minerals so called as protective food and are integral part of human diet. Keeping in view the value of these crops to human health these are recommended for consumption for good human health. According to National Research Council, Food and Nutrition Board, men and women between 23 to 50 years required to eat about 2800 and 2200 calories a day, respectively to maintain weight. The nutrient requirement varies with gender, age, height, weight, physical condition, activity level and climatic condition where they live *e.g.* dates supply 281 calories/100 g, avocado - 190 calories/100 g, banana - 95 calories/100 g, peach - 94 calories/100 g.

This book is enriched by including technical information related to production of horticulture crops such as importance and scope, classification, soil and climate, propagation methods and propagation structures, establishment of orchard, dormancy, juvenility, flowering, pollination, pollinator, fertilization, fruiting, unfruitfulness and fruit drop, training and pruning, methods of application of irrigation, manures, fertilizers and hormones *etc.*

I appreciate the efforts of authors Dr Mukesh Kumar and Dr Jeet Ram Sharma for preparing this comprehensive and informative book for the course Fundamentals of Horticulture as per syllabus approved by ICAR. I believe that this book will be helpful for the UG and PG students in Agriculture, Horticulture, Forestry and for competitive examinations like JRF/SRF/NET of agriculture/horticulture and its allied branches. It will also be helpful for researcher, teachers, extension workers and fruit growers to empower their knowledge.

Dr. Naresh Kaushik

Preface

The book Fundamentals of Horticulture is prepared as per the course curriculum approved by CCS HAU, Hisar, SAUs and ICAR. All the topics such as importance and scope, classification, soil and climate, planning and layout, dormancy, juvenility, flowering, pollination, reproduction, unfruitfulness, fruiting, fruit drop, propagation and propagation structures, application of growth regulators, irrigation, manures and fertilizers *etc.,* have been covered minutely. The topics are elaborated with pictorial representation for understanding of the concept.

We hope this book will be helpful to serve as a handy guide for students of graduation, post-graduation and preparation for JRF/SRF/NET in horticulture/ agriculture and other competitive examinations. It will be also helpful for researchers, extension workers, orchardist and others interested fellows in fruit culture.

For writing this book we have consulted several books, research papers, reports and web content from different institution sites in order to bring in the most recent and relevant information to readers. The efforts have been made to cover all the concepts on different topics but there are always chances of improvement in each and every document. The chemicals/ pesticides banned or may be banned in future by Government of India should not be recommended or used.

In spite of our best efforts, we do not claim this book is complete in all sense it may have some shortcomings, the authors will be really grateful to receive the suggestions and criticism from the readers for improvement and incorporation in forthcoming issues.

We are thankful to the University authority, various scientist, colleagues, students and others who helped directly or indirectly in preparation of this book. We also tribute to our lovely family for their inspiration and cooperation during preparation of this manuscript.

Mukesh Kumar

J. R. Sharma

Contents

Chapter 1

Horticulture: Its Definition and Branches, Importance and Scope of Horticultural Crops

History of Fruit Cultivation is as Old as Civilization

Fruits have their references in *Vedas, Purans, Upanishads etc.*

Oldest fruit plant - Date palm (reference dates back to 7000 B.C.)

Pomegranate (3500 B.C.)

Grapes (2440 B.C.)

Mango, Banana and Coconut (2000 B.C.)

Peach and Almond (1300 B.C.), Olive (100 B.C.)

First book exclusively on litchi cultivation was published in China in 1056 A.D.

Horticulture: The term is derived from the Latin word *'hortus'* – garden and *'cultura'* – cultivation. It deals with the cultivation of garden crops, which includes fruits, vegetables and ornamental crops.

Horticulture: It is a branch of agricultural sciences deals with intensively cultured plants directly used by people for food, medicinal purpose or aesthetic gratification.

Modern Horticulture: It is a branch of agricultural sciences which deals with the production, utilization improvement and post-harvest management of fruits, vegetables, ornamental plants, spices, plantation, medicinal and aromatic crops.

Branches of Horticulture

Pomology: Pomology is a branch of horticulture which deals with fruits. '*Poma*' is derived from Greek means fruits later subsequently transfer to '*Pome*' Latin word means fruits and '*logos*' mean study.

Branches of Pomology

a) **Basic Pomology**: Deals with the basic aspects of fruit production like methods of plantation, irrigation, training and pruning, water, nutrients and pest management, use of plant growth regulators.

b) **Commercial Pomology**: It deals with commercial fruits production technology.

c) **Systematic Pomology**: It deals with nomenclature, classification of fruit plants into kingdom, order, suborder, class, family, genus and species.

Olericulture: The '*oleris*' is derived from the Latin means pot herb and '*cultura'* means cultivation. Olericulture is branch of horticulture which deals with vegetables. Basically, olericulture is the art and science of the so-called truck-crops.

Floriculture/Landscaping: Floriculture is the science of flowers production. It deals with all the operations and practices used to take care of all such plants which are cultivated exclusively for their flowers and foliage. Landscape is the appearance of land as it appears to eyes. Landscape gardening is the art of beautifying a piece of land using garden design, methods and plant materials.

Facts about Fruit Production

- ☆ India is 7th largest country in the world with geographical area 328.73 m ha and 1st in population (140 crore), and 2nd in fruit production after China.
- ☆ Total arable land: 144 m ha.
- ☆ In India about 55-60 per cent population directly or indirectly depends on agriculture and its allied activities.
- ☆ India ranks 1st in productivity of grapes *i.e.* 20.81 MT/ha and 3rd in productivity of Banana *i.e.* 35.87 MT/ha.

Table 1.1: Position of India at World Level

Fruit Crops	*Rank at World Level*	*Share in World Production*
Mango/guava/mangosteen	1st	40.4 per cent
Papaya	1st	44.4 per cent
Banana	1st	25.7 per cent
Lemon, lime	1st	17.2 per cent

Total horticultural area in India : 28.04 m ha

Production : 347.18 m MT

Fruits

Area: 7.06 m ha

Production: 107.51 m MT

Productivity: 15.22 MT/ha

Plantation

Area: 4.29 m ha

Production: 15.76 m MT

Productivity: 3.67 MT/ha

Production share of crops in India among fruit crops

Banana: 32.1 per cent

Mango: 19.3 per cent

Citrus: 13.7 per cent

Papaya: 4.96 per cent

Guava: 5.14 per cent

Table 1.2: Leading Countries in Area and Production of Fruits

Countries	*Production (mMT)*	*Countries*	*Area (mh)*
China	272.08	China	16.59
India	90.89	India	6.98
Brazil	39.69	Brazil	2.33
USA	27.11	Philippines	1.68
Mexico	21.43	Spain	1.63

Table 1.3: Leading States in Area and Production of Horticultral Crops

States	*Production Share (per cent)*	*States*	*Area Share (per cent)*
Uttar Pradesh	12.70	Karnataka	9.74
Madhya Pradesh	10.17	Uttar Pradesh	8.72
West Bengal	9.56	Madhya Pradesh	8.60
Maharashtra	8.86	Maharashtra	8.53
Gujarat	7.56	West Bengal	7.05
Andhra Pradesh	7.49	Gujarat	6.79

Area and production of fruits in Haryana

Area: 69.14 (000'ha)

Production: 1010.49 (000'MT)

Productivity: 14.61 MT

Importance and Scope of Fruit Production

1. Economic Importance

A well maintained orchard brings better returns per unit area than field crops. Although initial expenses involved in establishment of an orchard are much more than raising field crops, but this can be compensated to large extent by taking inter crop during the initial years of plantation. High value crops with high potential of value addition make higher contribution to GDP *i.e.* 30 per cent from 14 per cent of total cropped area under horticultural crops. The export share of horticultural crops is 40 per cent of total agricultural output.

a) **High productivity:** Production per unit area from fruit crops is higher than other agricultural crops. The average yield of fruit crops like papaya, banana and grapes is 10 to 15 times more than that of agricultural crops.

b) **High net profit:** No doubt the establishment cost of new orchard is higher than field crops, but it is compensated by high production, its value, productivity and net profit *e.g.* the yield of wheat/ground nut/*ragi* is about 3.0 - 4.0 tones/ha valued Rs. 25000/- to 35,000/- as compared to production of grapes/mango/banana which is about 20-40 t/ha valued Rs.1.5-2.5 lakh.

c) **Provide input/raw material for agro based industries:** Fruit is one of the important inputs for the different fruit processing industries and other agro-industries –like canning and preservation of fresh fruits like aonla, bael, coir industries (coconut husk), pharmaceutical industry (aonla, bael, papaya, jamun), transportation and packaging industries (banana leaf and wood of other fruits) *etc.* Many horticultural produces and their byproducts are being used as raw materials for many industries, *e.g.* rubber, coir (coconut) and sago (tapioca) industries.

d) **Efficient utilization of resources:** All the natural as well as artificial farm resources may be utilized throughout the year because fruit crops are perennial in nature so farm labour, machinery, land, water are required throughout the year as compared to field crops.

f) **Utilization of waste and barren lands for production:** Most of the fruit crops are perennial in nature and require good quality water and soil for production, however some arid and semi-arid fruit crops are of hardy nature *e.g.* ber, aonla, phalsa, jamun, cashew, custard apple *etc.*, which can be grown successfully on poor shallow, undulated soils considered unsuitable for growing other crops.

g) **Foreign exchange:** Export of fresh fruits, vegetables and their processed products are in great demand in the foreign countries and thus are source of earning foreign exchange *e.g.* fresh fruits and vegetables – apple, mango, grapes, mushroom, onion, potato *etc.*, flowers – cut flowers (roses), potted and foliage plants, plantation – cashewnut, tea, coffee, coconut, arecanut *etc.*

2. Important for Nutritional Security/Protective Food

Horticultural crops are rich source of vitamins and minerals so called as protective food. Keeping in view the value of these crops to human health these are recommended for good human health 230 g of fruits and 400 g of vegetables per capita per day.

$$\textbf{\textit{Per capita availability of fruits (g)}} =$$

$$\frac{Total\ Production\ (g) - (Post\ harvest\ Losses + Export)}{Total\ Population\ x\ 365}$$

a) Carbohydrates Fats and Protein

Carbohydrates and fats provide energy to human body while protein and fat are essential and play an important role in the growth and development of the human being.

Table 1.4: Composition of Carbohydrates, Fats and Protein in Fruits

Fruit Crops	*Carbohydrates (per cent)*	*Fat (per cent)*	*Protein (per cent)*
Almond	10.5	58.7 (2nd)	20.8 (2nd)
Apricot (dry)	72.8 (2nd) *	0.7	1.7
Avocado	0.8	22.8	1.7
Banana	36.4	0.2	1.3
Bael	30.6	0.2	1.3
Ber	13.6	0.1	0.9
Cashewnut	22.3	46.9 (3rd)	21.2 (1st)
Dates	67.3 (3rd)	0.2	3.0
Karonda	67.1 (4th)	9.6	2.3
Raisins (dried grapes)	77.3 (1st)	0.2	2.0
Walnut	11.0	64.5 (1st)	15.6 (3rd)

* Ranks

b) Calorific Value

According to National Research Council, Food and nutrition board, men and women aged between 23 to 50 years eat about 2800 and 2200 calories a day, respectively to maintain weight. The nutrient requirement varies with gender, age, height, weight, physical condition, activity level and climatic condition where they live *e.g.* dates supply 281 calories/100 g, avocado - 190 calories/100 g, banana - 95 calories/100 g, peach - 94 calories/100 g.

c) Vitamins

Vitamins are important constituents of fruits and vegetables and are essential part of human diet. These required in very little amount but essential for maintenance of health. The deficiency of any vitamin may lead to disease or disorder. Following vitamins are supply by fruits and vegetables.

i) **Vitamin-A (carotene):** It is essential for normal growth, reproduction and maintenance of health and vigour. It protects against cold, influenza and prevents night blindness. Its deficiency results in cessation of growth in young children, night blindness, drying up of tear glands in the eyes, eruption of skin (rashes on the skin) and brittleness of the teeth.

Sources: Mango (4800 IU/100 g), papaya (2020 IU/100 g), dates, jackfruit, walnut *etc.*

ii) **Vitamin B_1 (Thiamine):** Tones the nervous system and helps in proper functioning of digestive tract. Its deficiency cause beri-beri, paralysis, loss of sensitivity of skin, enlargement of heart, loss of appetite, loss of weight and fall in body temperature. (one IU biologically equivalent to 0.3 mcg retinol, mcg-micrograms).

Sources: Cashew nut (630 mg/100 g), walnut (450 mg/100 g), dry apricot (217 mg/100 g), orange, pineapple, jackfruit, almond, banana *etc.*

iii) **Vitamin B_2 (Riboflavin):** It is required for body growth and health of the skin. The deficiency of this vitamin causes sore throat, anorexia cataract, and loss of appetite, body weight and also development of swollen nose.

Sources: Bael (1191 mg/100 g), papaya (250 mg/100 g), litchi (122 mg/100 g), banana, apricot, pomegranate, pear *etc.*

iv) **Vitamin -C (Ascorbic Acid):** It promotes general health and healthy gums, prevents scurvy disease which is characterized by pain in the joints and swelling of limbs (rheumatism), bleeding of gums, tooth decay and keeps the blood vessels in good condition.

Sources: Barbados cherry (1500-4000 mg/100 g), aonla (600-1200 mg/100 g), guava (220-300 mg/100 g), ber (120 mg/100 g), citrus (50-60 mg/100 g), strawberry, pineapple *etc.*

d) Minerals

A large number of minerals are present in human body. Bones and teeth are mainly made up of calcium, magnesium and phosphorus. Iron is an important constituent of blood whereas iodine regulates the functioning of hormones (thyroxine). Minerals like zinc, molybdenum and manganese are activators of certain enzymes. In addition to these, many of these elements are also required for adequate growth of tissues.

Table 1.5: Minerals Composition in Fruits

Fruit Crops	*Calcium (per cent)*	*Phosphorus (per cent)*	*Iron (mg/100 g)*
Almond	0.23 (1st)	0.39	3.47
Aonla	0.05	0.02	1.20
Apple	0.01	0.01	1.68
Ber	0.03	0.04	1.14
Cashewnut	0.05	0.45 (1st)	5.00 (3rd)
Dates	0.07	0.08	10.60 (2nd)
Grapes	0.005	0.02	0.40
Guava	0.01	0.04	1.04
Karonda	0.16 (3rd)	0.06	39.70 (1st)
Mango	0.007	0.02	0.25
Papaya	0.01	0.01	0.27
Peach	0.006	0.03	1.70
Raisins	0.10	0.08	4.00
Walnut	0.10	0.38 (2nd)	4.80
Litchi	0.21 (2nd)	0.30 (3rd)	1.00

e) Enzymes

Enzymes required for different metabolic activities in body.

Sources: Papaya-papain and pineapple-bromelain.

f) Fibre and Roughages (Cellulose and pectin)

These are required for digestion and prevention of constipation.

Sources: Fruits are low in fibre content. However, guava (6.90 per cent), anola (5.20 per cent) and pomegranate (5.10 per cent) are better sources compared to other fruits. Leafy vegetables are rich in fibre content.

g) Medicinal values

Some fruits like aonla, bael, pomegranate, jamun, karonda, guava, kokum, ber *etc.*, having medicinal value and these are recommended for flowing disorders:

- Papaya is rich source of vitamin A, it helps to reduce night blindness.
- Citrus juice reduces acute diarrhea.
- Aonla is rich in vitamins C and an important constituent in *triphala* (*chawanprash*)- helps in digestion.

3. Social and Religious Importance

Horticulture is a part of civilization. Fruits are being used for worship of Gods. Bael leaves are used to worship Lord Shiva. Tying of mango leaves at the entrance of house is considered auspicious. Due to higher returns per unit area from the horticultural crop, the person engaged in this activity have higher social status.

4. Employment Opportunity

Horticultural crops provide employment for small farmers and agricultural labour throughout the year. One hectare of fruit production generates 860 man-days per annum as against 143 man-days for cereal crops and the crops like grapes, banana and pineapple need 1000- 2500 man-days per annum.

5. Other Importance

- ☆ Horticultural trees work in near resemblance to forest trees in maintaining ecosphere.
- ☆ Fruit crops provide better micro climate.
- ☆ These are perennial in nature so provides shelter to birds, small animals, reptiles and other microbes and add diversity in the ecology.
- ☆ These provides impetus and calm environment for thinkers, writers, poets and analysts to convert their thinking in words.
- ☆ Adds value to the all living entity.
- ☆ In addition to the above these are also beneficial like forest plants.

We can say the horticultural crops are better for health, wealth, happiness and hygiene.

Scope of Horticulture

In brief it can be stated that horticulture has great scope for the following reasons:

- ☆ Horticultural crops are grown in different climatic zones so variability can be exploited in different agro-climatic conditions.
- ☆ It is helpful to meet out the nutrition demand and other needs of growing population for fruits, vegetables, flowers, spices and beverages *etc.*
- ☆ To meet out the requirement of raw material for different processing industries.
- ☆ To substitute import and earning foreign exchange by increasing export.
- ☆ More production, productivity and market value improve social status of the farmers and also decrease unemployment problem by engaging more labour in horticultural crops as compared to field crops.
- ☆ Helps to increase greenery, carbon sequestration, maintain biodiversity and protect environment.

Constraints

A) Production Constraints

- ☆ Inadequate quality planting material.
- ☆ Low density plantation.
- ☆ Lack of knowledge.
- ☆ Untrained labour.
- ☆ Poor tree canopy management/tree architecture.
- ☆ Small land holding/decreasing land holding.
- ☆ Long gestation period.
- ☆ Pre-dominated old and/senile unproductive orchards.
- ☆ Poor quality produce.
- ☆ Non-adoption of appropriate agro-techniques and good management practices.

B) Protection Constraints

- ☆ Indiscriminate use of pesticides and fungicides for immediate control of major pest and diseases and lack of IPM.
- ☆ Few national problems in fruit crops are mango malformation, guava wilt, citrus decline, citrus granulation, biennial bearing *etc.*

C) Postharvest Handling Constraints

- ☆ Early harvesting of fruits before attaining maturity.
- ☆ Lack of basic post-harvest handling infrastructure facilities.
- ☆ Non-adoption of post-harvest technologies.
- ☆ Absence of pre-cooling, cold chain and cold storage.
- ☆ Non-availability of processing specific varieties.

d) Policy Support

- ☆ Lack of suitable laws in nursery act and contract farming.
- ☆ Lack of credit support to farmers.
- ☆ Lack of trainings to farmers/growers.
- ☆ Formulation of standards for planting materials, produce quality, photosanitary, packaging standards as well as standards for pesticide residues.
- ☆ Preparation of standards on quality system *viz.*, HACCP (hazard analysis critical control points), GMP (good manufacturing practices), good hygiene and sanitation practices.

☆ Lack of market facility, market intelligence and market information services (MIS).

Strategies

1. Increase availability of good quality seeds and planting material.
2. Rejuvenation of old orchard.
3. Optimum utilization of land space by adopting high density planting.
4. Soil health management/Nutrient management to be adopted.
5. Leaf nutrient analysis techniques to be followed (DRIS- Diagnostic and recommendation integrated system).
6. Efficient management of inputs like nutrients, irrigation water (through use of plastic mulching and drip irrigation), green houses, low tunnels, shading nuts, film for drying produce, propagation in nurseries and water conservation through pond lining.
7. Adoption of biocontrol measures for major insect pest and diseases to reduce pesticide toxicity. Promotion for organic farming should be given due attention.
8. Use of growth promoting chemicals for increasing productivity *e.g.* use of gibberellins in grape and paclobutrazol (Cultar) in mango.
9. Practice of precision farming in horticulture. Precision farming is efficient management of resources through location specific high technology interventions.
10. Proper packaging and storage to avoid post-harvest losses.
11. Development/utilization of cold chain facility.
12. Establishment of processing industries in intensive fruit growing area.
13. Make fruit production-oriented credit policies and timely release of assistance as per the seasonal requirement.
14. Formulation of standards on quality, planting material, phytosanitory as well as standards for pesticides residue in fruit crops.
15. Training to growers, single window system to solve growers' problems.

Hi-tech Horticulture

Hi-tech technologies can be defined as those, which are modern, less environment dependent, capital intensive and have the capacity to improve the productivity and quality. These includes use of GM (genetically modified) crop/varieties, HDP (high density planting), micro propagation, integrated nutrient, water and pest management, use of plastics mulching, precision farming, protected cultivation, organic farming, use of modern immunodiagnostic techniques for quick detection of viral disease and hi-tech post-harvest technologies including cold chain.

List of NRCs/Institutes/Directorates/Universities exclusively working on Horticulture (source: www.icar.org.in)

National Research Centers (NRCs)

- ICAR-National Research Centre for Banana, Trichi, Tamil Nadu
- ICAR-National Research Centre for Grapes, Pune, Maharashtra
- ICAR-National Research Centre for Litchi, Muzaffarpur, Bihar
- ICAR-National Research Centre for Pomegranate, Solapur, Maharashtra
- ICAR-National Research Centre on Orchids, Pakyong, Sikkim
- ICAR-National Research Centre on Plant Biotechnology, New Delhi
- ICAR-National Research Centre on Seed Spices, Ajmer, Rajasthan
- Central Institute of Horticulture, Medziphema, Nagaland

ICAR Institutes

- ICAR-Central Institute for Arid Horticulture, Bikaner, Rajasthan
- ICAR-Central Institute of Sub Tropical Horticulture, Lucknow, UP
- ICAR-Central Institute of Temperate Horticulture, Srinagar, J & K
- ICAR-Central Institute on Post harvest Engineering and Technology, Ludhiana, Punjab
- ICAR-Central Plantation Crops Research Institute, Kasaragot Kerala
- ICAR-Central Potato Research Institute, Shimla, Himachal Pradesh
- ICAR-Central Tobacco Research Institute, Rajahmundry, Andhra Pradesh
- ICAR-Central Tuber Crops Research Institute, Trivandrum, Kerala
- ICAR-Indian Institute of Spices Research, Kozhikode, Kerala
- ICAR-Indian Institute of Vegetable Research, Varanasi, UP
- ICAR- Indian Institute of Oil Palm Research, Pedavegi, West Godawari, Andhra Pradesh
- ICAR-Central Citrus Research Institute, Nagpur, Maharashtra

Directorates

- ICAR-Directorate of Cashew Research, Puttur, Karnataka
- ICAR-Directorate of Medicinal and Aromatic Plants Research, Anand, Gujarat
- ICAR-Directorate of Floricultural Research, Pune, Maharashtra

Universities

- ☆ Maharana Pratap Horticulture University, Karnal, Haryana
- ☆ Dr. Yaswant Singh Parmar University of Horticulture and Forestry, Solan, Himachal Pradesh
- ☆ University of Horticulture Science, Bagalkot, Karnataka
- ☆ University of Agriculture and Horticulture Sciences, Shimoga, Karnataka
- ☆ Sri Konda Laxman Telangana State Horticultural University, Hyderabad, Telangana
- ☆ VCSG Uttarakhand University of Horticulture and Forestry, Bharsar, Uttarakhand

Sources of Database

www.nhb.gov.in

https://www.fao.org/faostat/en/

https://apeda.gov.in/

https://agricoop.nic.in/en/statistics/publication-reports

www.hortharyana.gov.in

https://www.indiastat.com/Home/Login

https://icar.org.in

Chapter 2

Botanical Classification

Classification of Fruits

Arrangement of individuals according to characters for systematic study. It helps the pomologist, students and farmers to classify the crops. It helps:

- ☆ To identify and naming nomenclature of crop.
- ☆ To study the close relationship between species.
- ☆ To know their crossing and hybrids behavior.
- ☆ To know their compatibility and inter grafting ability.
- ☆ To know their adaptability to soil and climate.

Classification of Fruits Based on Climatic Adaptability

Fruit crops are classified into following three categories on the basis of the climate:

Temperate Fruits

The temperature remains less than 4°C and generally snowfall takes place during winter and in summer temperature ranged from 10-14°C. This zone lies at a height between 1800-3500 m from mean sea level. The plants generally require a specific chilling temperature and are deciduous in nature *e.g.* apple, almond, peach, pear, plum, strawberry, apricot, persimmon, cherimoya, pecan nut, walnut, hazelnut, cherry, pistachio and kiwi fruit *etc.*

Sub-tropical Fruits

- ☆ The area between tropical and temperate conditions is called subtropical.

The temperature in subtropical area ranges from 25 to 30°C and height from mean sea level varies from 900 to 1800 m.

- Plants in these conditions are deciduous or evergreen and are usually unable to withstand frost but can tolerate low temperature. However, some plants require chilling hours for flower bud differentiation. Example; grape, durian, citrus, jackfruit *etc.*
- These plants are adaptive to fluctuations in photoperiod.

Tropical Fruits

- Fruit crops in tropical conditions are evergreen and are sensitive to low temperature (cold).
- Temperature in these conditions ranges from 22 to 27°C and height from mean sea level varies from 300 to 900 m. Plants require moist warm climate but can also withstand dry weather to some extent *e.g.* papaya, mango, banana, sapota *etc.*
- Fluctuation in photoperiod and temperature is very less.

Classification Based on Nature of Growth (Nature of Stem or Height)

Herbaceous: These are non-woody plants having soft and green stem *e.g.* banana, pineapple *etc.*

Shurbaceous: These are small to medium sized perennial plants up to a height of about 3 m *e.g.* karonda, phalsa, pomegranate *etc.*

Woody: These plants have hard stem and having height above 3 m *e.g.* mango, ber, sapota, jamun, guava, apple, peach, pear *etc.*

Classification Based on Continuation of Growth

Evergreen: Plants retain leaves throughout the year *e.g.* mango, citrus, litchi, sapota *etc.*

Deciduous: Plants lose their leaves once in a year *e.g.* apple, pear, peach, plum, apricot *etc.*

Classification Based on Bearing Habit

On the basis of age and location of the shoot to bear flower or fruit these are classified into six categories. On the basis of these the time and intensity of cultural operations like training, pruning and heading back are decided.

- Terminal bearing on current season growth: Jackfruit, pecan, loquat *etc.*
- Terminal bearing on old season growth: Mango, Pineapple, litchi, banana *etc.*
- Axillary bearing on current season growth: Papaya, orange, coconut, guava, passion fruit *etc.*

- ☆ Axillary bearing on old season growth: Apple, custard apple, pear, peach, plum *etc.*
- ☆ Mixed bearing: Pomegranate, citrus, carambola *etc.*

Table 2.1: Fruits Classified on the Basis of Ripening behaviour/Rate of Respiration

Climacteric	*Non-climacteric*
Mango, sapota, banana, peach, guava, fig, papaya, annona, pear, plum, apple, tomato	Grape, citrus, pomegranate, litchi, pineapple, ber, cashew, jamun, cherry, strawberry, cucumber

(Rate of respiration and production of ethylene is at very fast rate in case of climacteric fruits as compared to non-climacteric fruits)

Table 2.2: Classification on the Basis of Response to Photoperiod

Long Day	*Short Day*	*Day Neutral Plant*
Passion fruit, apple, banana	Strawberry, coffee, pineapple	Guava, papaya, marigold

Table 2.3: Classification on the Basis of Salt Tolerance

Highly Tolerant	*Medium Tolerant*	*Highly Sensitive*
Date palm, ber, aonla, pomegranate, khirni, coconut	Jamun, cashew, guava, fig, phalsa	Papaya, mango, apple, pear, citrus, strawberry

Table 2.4: Classification on the Basis of Acid Tolerance

Highly Tolerant	*Medium Tolerant*
Strawberry, raspberry, fig, bael, plum	Pineapple, avocado, litchi

Classification of Fruits on the Basis of Edible Parts

Fruits: Mature or ripened ovary

Fruits:

i) Pericarp (thick and thin): Epicarp, mesocarp, endocarp.

ii) Seed

(a) Parthenocarpy: Development of seed without proper fertilization. Such fruits are seedless or immature seeds *e.g.* banana, grapes, orange, apple *etc.*

(b) True fruits (Eucarp): When ovary grow into a fruit.

(c) False fruit (Pseudocarp): When floral parts like thalamus or calyx grow in the form of fruit *e.g.* apple, pear, cashewnut.

Classification Based on Fruit Morphology

1. Simple Fruit

i) Berry: Pericarp is divided into exo (outermost layer), meso (juicy and pulpy thick layer) and endocarp (thin and papery) *e.g.* grape, guava, banana, papaya, date, sapota.

ii) Modified berry:

a) **Stone/drupe:** Pericarp is divided into distinct layer epicarp (thick), mesocarp (pulpy) and endocarp (hard and woody containing single seed) *e.g.* plum, apricot, peaches, mango, coconut, jamun.

b) **Pome:** False fruit developed from ovary along with adjacent flower fleshy thalamus *e.g.* apple, pear, loquat, quince.

c) **Hesperidium:** Leathery and granular pericarp, locules of the fruits contain juicy sacs developed from placental hair of endocarp *e.g.* oranges.

d) **Nut:** Pericarp become thick and hard *e.g.* litchi, rambutan.

e) **Capsule:** Developed from two or more united carpels and fruit are many seeded *e.g.* aonla, carambola.

f) **Balausta:** False berry in which syncarpous pistil consisting of two rows of fused carpels, one above the other. The whole fruit is covered by a hard rind made up of exocarp and part of mesocarp. Mesocarp also forms plate like infolding. The endocarp is papery and covers the individual groups of seeds *e.g.* pomegranate.

g) **Pepo**: Faba berry in which the rind or exocarp does not separate from mesocarp. Similarly, seeds do not separate from the placenta which are inverted to form septa *e.g.* gourd, cucumber, melon, pumpkin *etc.* Rinds may or may not be edible. Normally endocarp, mesocarp as well as placenta (except melon) are edible *e.g.* watermelon.

h) **Amphisarca**: Multilocular and many seeded superior berries in which outer part of pericarp or epicarp is woody, mesocarp is fibrous while endocarp and placental form the edible pulp *e.g.* bael, wood apple.

2. Aggregate Fruits (etaerio – aggregate of simple fruits borne by a simple flower)

Fruit developed as aggregation of fruits developed from several ovaries or collection of several simple fruits from an apocarpous (free carpels) pistil of flower. Important aggregate fruits are as follow:

i) Etaerio of achenes: It is an aggregate fruit in which the individual fruitlets are achenes (dry single seeded indehiscent fruitlets in each of which the seed is attached to the dry pericarp at one point *e.g.* strawberry has brownish achenes embedded over the surface of fleshy and edible thalamus).

ii) Etaerio of samaras: *e.g.* ailanthus

iii) Etaerio of follicles: *e.g.* larkspur, calotropis

iv) Etaerio of berries: *e.g.* custard apple - in custard apple (sitaphal) individual berries are fused together around an elongated central axis. They are demarcated from outer side by hexagonal areas. Edible parts are the juicy mesocarp of individual berries.

v) Etaerio of drupes: *e.g.* raspberry, blackberry.

3. Composite or Multiple Fruits

It is a group of fruitlets which develop from different flowers of an inflorescence often fused with one another. These are of two types:

i) Sorosis: Composite fruit develop from either a spike of female catkin *e.g.* mulberry, pineapple, jackfruit.

Mulberry- develop from a female catkin. Ovaries of individual flowers develop into minute seedless fruitlets. Their perianth lobes (4 number) become fleshy and juicy. Whole fruit is edible.

ii) Syconus: It develops from hypanthodium type of inflorescence *e.g.* peepal, banyan, fig (all *Ficus* group). The flask shaped receptacle becomes fleshy. The enclosed female flowers may produce small achene like fruitlets. Syconus has a small sore protected by scale leaves.

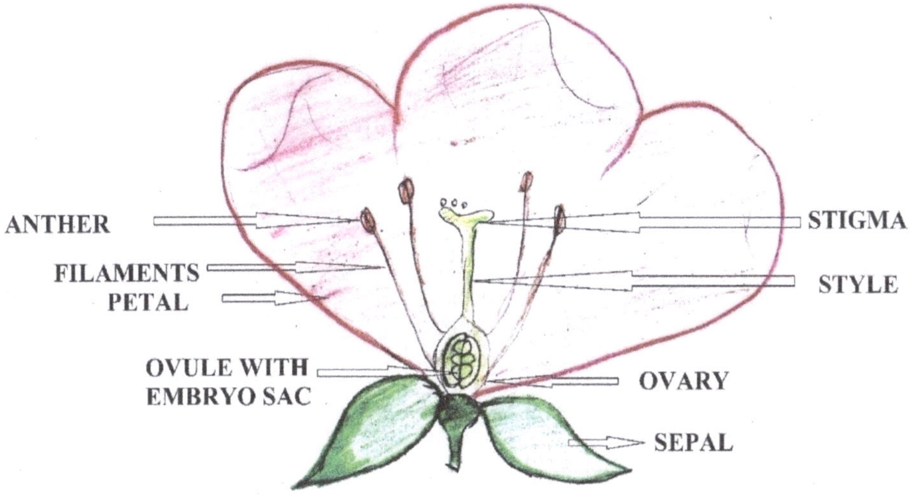

Fig 2.1: Parts of Flower

Female reproductive organs in flower

i) Apocarpous: In a compound pistil, the carpels may be free with as many ovaries as the number of carpels *e.g.* champa.

Table 2.5: Botanical Classification along with Type of Fruits, Edible Part and Ploidy Level

Common Name	*Botanical Name*	*Family*	*Type of Fruit*	*Edible Part*	*Ploidy No. (2n)*
Monocotyledon					
Arecanut	*Areca catechu*	Arecaceae/Palmae			32
Banana	*Musa paradisiaca*	Musaceae	Berry	Mesocarp and Endocarp	22,33,44
Coconut	*Cocus nucifera*	Arecaceae/Palmae	Drupe	Endosperm	32
Date palm	*Phoenix dactylifera*	Arecaceae/Palmae	Drupe	Pericarp	36
Oil Palm	*Ealias guinensis*	Arecaceae/Palmae	Drupe	Kernel	32
Pineapple	*Ananas comusus*	Bromeliaceae	Sorosis	Bracts/perianth	50
Dicotyledon					
Almond	*Prunus communis*	Rosaceae	Drupe	Seed	16
Aonla	*Phyllanthus emblica/ Emblica officinalis*	Phyllanthaceae Euphorbiaceae	Capsule	Mesocarp Endocarp	28
Apple	*Malus domestica*	Rosaceae	Pome	Fleshy thalamus	34
Apricot	*Prunus armeniaca*	Rosaceae	Drupe	Mesocarp, Endocarp	16
Avocado	*Perisa americana*	Lauraceae	Berry	Pericarp	24,36,48
Bael	*Aegle marmelos*	Rutaceae	Amphisarca	Succulent placenta	18
Ber	*Ziziphus mauritiana*	Rhamnaceae	Drupe	Pericarp	48(4x)
Bread fruit	*Artocarpus baltilis*	Moraceae	Sorosis	Fleshy receptacle	56 (4x)
Cacao	*Theobroma cocoa*	Sterculiaceae	Drupe	Beans	20
Carambola/Star fruit	*Averrhoa carambola*	Oxalidaceae	Capsule		22
Cashew nut	*Anacardium occidentalle*	Anacardiaceae	Nut	Fleshy peduncle, cotyledon and seed	42
Coffee	*Coffea robusta*	Rubiaceae	Drupe	Seed	22

Common Name	*Botanical Name*	*Family*	*Type of Fruit*	*Edible Part*	*Ploidy No. (2n)*
Custardapple	*Annona squamosa*	Annonaceae	Aggregate of berry (Etaerio)	Fleshy pericarp of berries	14
Durian	*Durio zibethinus*	Bombaceaceae	Berry	Arils	56
Fig	*Ficus carica*	Moraceae	syconus	Fleshy receptacle	56
Grape	*Vitis vinifera*	Vitaceae	Berry	Pericarp, placentae	38
Grapefruit	*Citrus paradisi*	Rutaceae	Hesperidium	Juicy placental hairs	18
Guava	*Psidium guajava*	Myrtaceae	Berry	Thalamus and pericarp	22
Jackfruit	*Atrocarpus bheterophyllus*	Moraceae	Sorosis	Bracts/perianth and seed	56
Jamun	*Syzygium cumini*	Myrtaceae	Drupe	Mesocarp Epicarp	40
Karonda	*Carissa carandus*	Apocyanaceae	Berry	Mesocarp Epicarp	22
Kagzi lime	*Citrus aurantifolia*	Rutaceae	Hesperidium	Juicy placental hairs	18
Kiwi fruit	*Actinidia chinensis*	Actinidaceae	Berry	Fleshy mesocar & Endocarp	58
Lasora	*Cordiya mixa*	Boraginaceae	Drupe	Pericarp	—
Lemon	*Citrus limon*	Rutaceae	Hesperidium	Juicy placental hairs	18
Litchi	*Litchi chinensis*	Sapindaceae	Nut	Mesocarp	28, 30
Loquat	*Eriobotrya japonica*	Rosaceae	Pome	Fleshy thalamus	34
Mandarin	*Citrus reticulata*	Rutaceae	Hesperidium	Juicy placental hairs	18
Mango	*Mangifera indica*	Anacardiaceae	Drupe	Mesocarp	40
Mongosteen	*Garcinia mangostana*	Guttiferae	Berry	Arils	74-110
Mulberry	*Morus alba*	Moraceae	sorosis	Perianth	308
Olive	*Olea europaea*	Oleaceae	Drupe	Mesocarp Epicarp	46

Common Name	*Botanical Name*	*Family*	*Type of Fruit*	*Edible Part*	*Ploidy No. (2n)*
Papaya	*Carica papaya*	Caricaceae	Berry	Mesocarp	18
Passion fruit	*Passiflora edulis*	Passifloraceae	Berry	Mesocarp	18
Peach	*Prunus persica*	Rosaceae	Drupe	Mesocarp Epicarp	16
Pear	*Pyrus communis*	Rosaceae	Pome	Fleshy thalamus	34
Pecan nut	*Carya illinoensis*	Juglandaceae	drupe	Seed	32
Persimmon	*Diospyros kaki*	Ebenaceae	Berry	Mesocarp Epicarp	90
Phalsa	*Grewia subinaequalis*	Tiliaceae	Drupe	Mesocarp Epicarp	36
Pistachionut	*Pistachia vera*	Anacardiaceae	Nut	Cotyledon	30
Plum	*Prunus domestica*	Rosaceae	Drupe	Mesocarp Epicarp	32
Pomegranate	*Punica granatum*	Punicaceae	Balausta	Juicy seed coat	18
Roseapple	*Syzygium jambos*	Myrtaceae	Drupe	Mesocarp	28, 33
Rumbutan	*Nephelium lappaceum*	Sapindaceae	Berry		32
Sapota	*Manilkara achras (Achras zapota)*	Sapotaceae	Berry	Mesocarp	26
Strawberry	*Fragaria ananassa*	Rosaceae	Etaerio of achenes	Fleshy thalamus	56
Sweet cherry	*Prunus avium*	Rosaceae	Drupe	Mesocarp	16
Sweet orange	*Citrus sinensis*	Rutaceae	Hesperidium	Juicy placental hairs	18
Tamarind	*Tamarindus indica*	Leguminoceae	Pod	Mesocarp	24
Tea	*Camellia sinensis*	Theaceae or Camaliaceae		Seed	30
Walnut	*Juglans regia*	Juglandaceae	drupe	Seed	32
West Indian cherry	*Malphigia puncifolia*	Malphigiaceae	Drupe/stone		20
Wood apple	*Feronia limonia*	Rutaceae	Amphisarca	Succulent placentae	18

(ii) Syncarpous: In a compound or polycarpellary pistil, carpels may be united into one ovary. This condition is more common *e.g.* guava

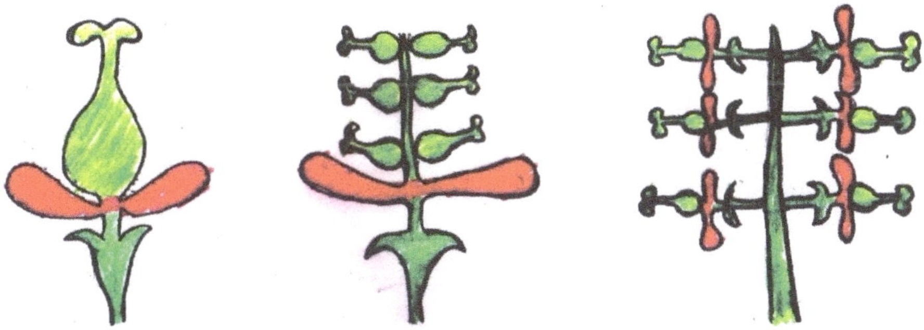

Simple Fruit **Aggregate Fruit** **Multiple Fruit**

Fig 2.2: Botanical Classification of Fruits

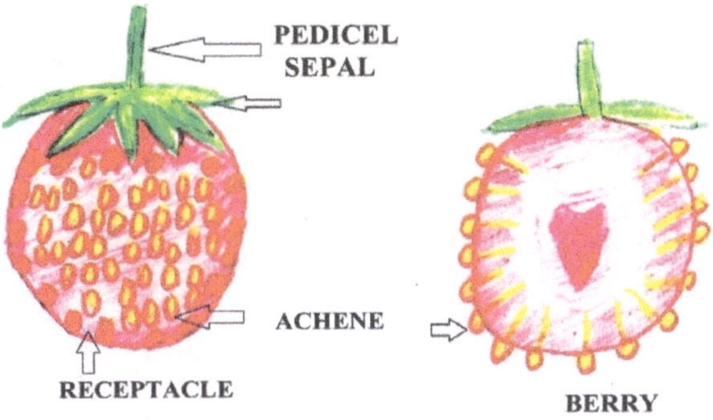

Fig 2.3: Berry (Strawberry)

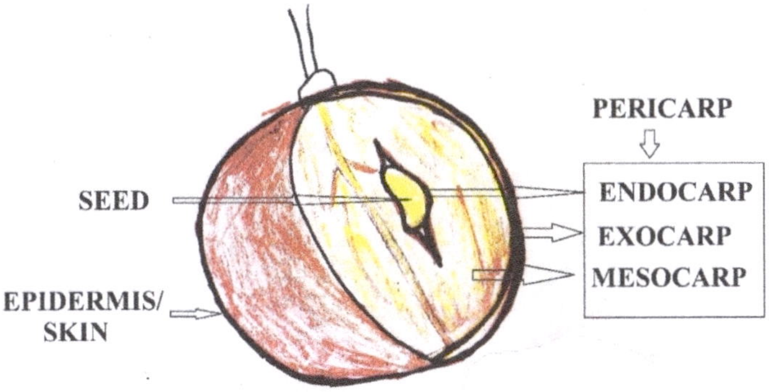

Fig 2.3 (a): Berry

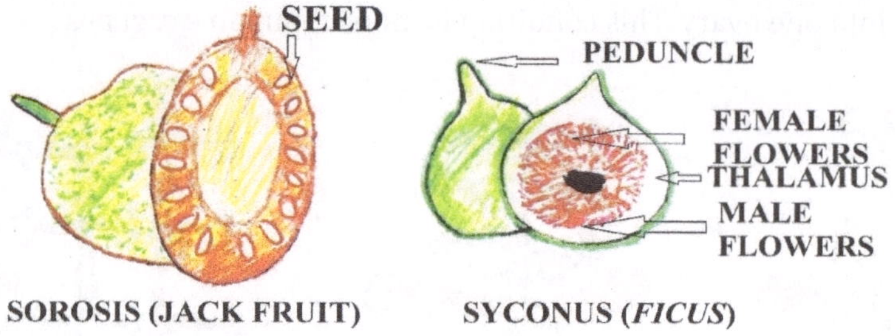

Fig. 2.4: Multiple or Composite Fruit

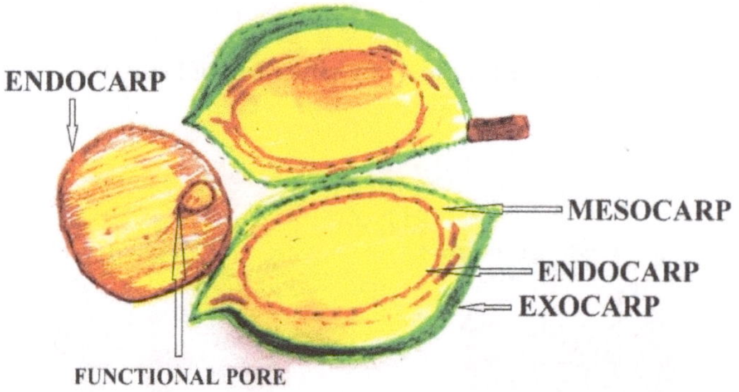

Fig 2.5: Drupe Fruit (Coconut)

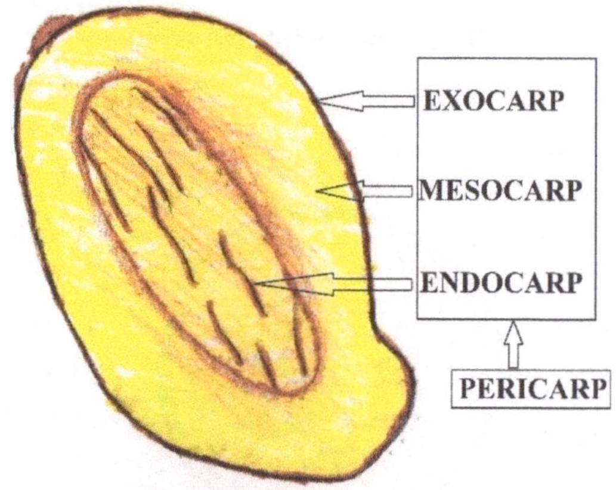

Fig 2.5 (a): Drupe Fruit (Mango)

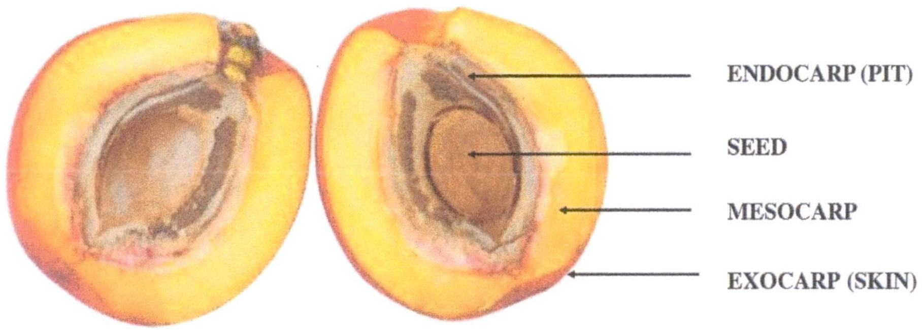

Fig. 2.5(b): Drupe (Fleshy Fruit with a Stony Endocarp)

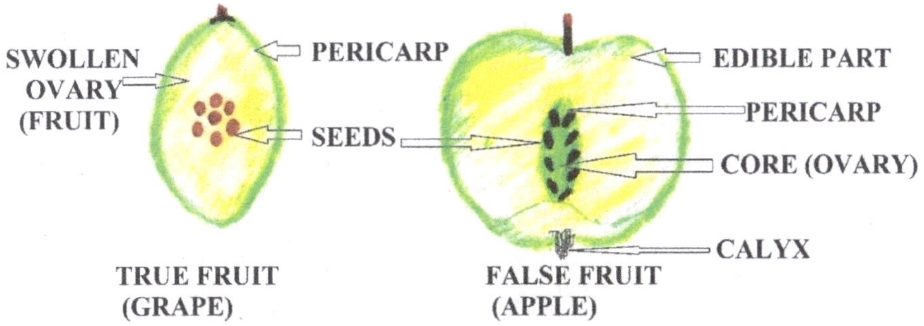

Fig 2.6: True Fruit and False Fruit

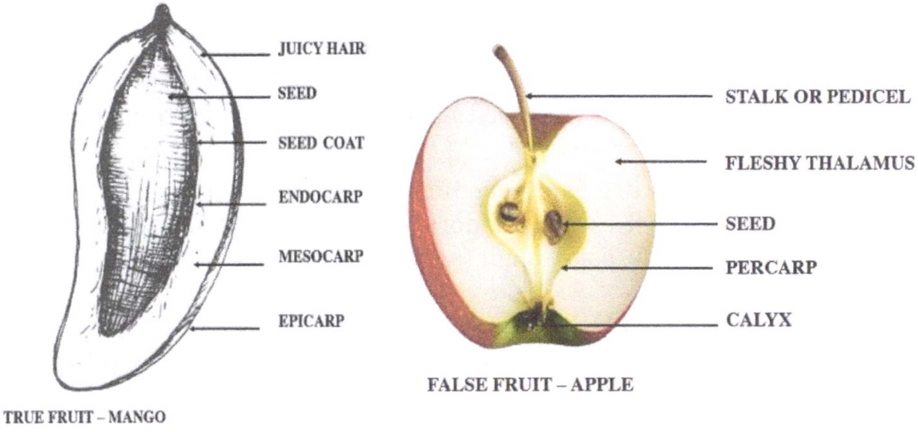

Fig 2.6(a): True Fruit and False Fruit

Types of Fruits

i) Simple fruits: Formed from flower having single pistil.
ii) Aggregate fruits: Flower having poly carpellary apocarpous gynoecium.
iii) Composite fruits: Fruits from whole inflorescence.

Simple Fruits

A fruit developed from a single flower possessing one or more syncarpous carpel. It can be dry or succulent.

Dry Fruits

i) Indehiscent fruits: These are single seeded indehiscent/pericarp fruit. The Fruit doesn't split or open at maturity to shed their seeds.
ii) Capsular: Pericarp splits open at maturity and the seeds are released. Many seeded dehiscent.
iii) Schizocarpic: It is developed from flowers having two or more united carpels which split into indehiscent one seeded part is called mericarp. Many seed breaking up into single seeded parts.

i) Indehiscent Fruits

Achenial fruits, dry and simple fruits

a) Achene: A small, dry one seeded fruit which develops from a single carpel. The pericarp of fruit does not have seed coat *e.g.* four O' clock plant, hogweed *etc.*
b) Caryopsis (grains): This is very small, dry one seeded fruit which develops from a single pistil with the pericarp fused with seed coat *e.g.* maize, wheat, rice, oat (Gramineae family).
c) Cypsela: This is dry one seeded fruit which develops form an inferior bicarpellary ovary *e.g.* marigold, sunflower.
d) Samara: The pericarp of single seeded indehiscent dry fruit bears membranous out growth called wings for wind dispersal *e.g.* ulmus (elm), holoptelea (Indian elm).
e) Nuts: This is a dry one seeded fruit which develops from a superior syncarpous pistil with the pericarp hard and woody *e.g.* almond, cashewnut, litchi, oak, walnut (earlier considered as drupe).

ii) Capsular (dehiscent) Fruits

Many seeded, dry, simple, dehiscent fruit in which pericarp splits open to expose seed.

a) Pod or legumes: These are dry one chambered fruit which develops from a simple pistil and dehiscing by both the margins *e.g.* pea, (Leguminosae-family).

b) Follicle: It is similar to legume but dehisces by only one suture *e.g.* larkspur.

c) Siliqua: This is a dry, long, narrow two chambered fruit developing from a bicarpellary pistil with two parietal placentae *e.g.* mustard, radish.

d) Silicula: A short broad and flat siliqua *e.g.* candytuft, shepherd purse.

e) Capsule: Dry, one-to-many chambered fruit which develops from a syncarpous pistil and dehiscing in various ways *e.g.* cotton, lady finger, linseed.

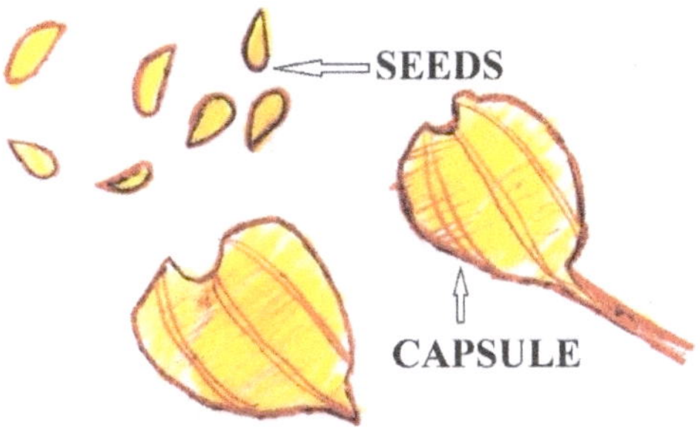

Fig 2.7: Capsule

iii) Schizocarpic Fruits (splitting)

Many seeded dry and simple fruit which break up into single seeded parts. The dehiscent single seeded parts are called as mericarp while the dehiscent ones are termed as cocci (singular coccus).

a) Lomentum: The fruit is constricted in between the seeds. It is either indehiscent or break up into single seeded parts. These are two types:

Lomentaceous siliqua: Fruit is internally like a siliqua but is constricted in between the seeds *e.g.* indehiscent fruit of radish.

Lomentaceous pod: When the pod is constricted or partitioned between the seed into a number of seeded compartments *e.g.* groundnut, (indehiscent), Acacia (cocci).

b) Cremocarp: This is dry two chambered inferior fruit splitting into two indehiscent one seeded piece called mesocarp *e.g.* coriander, fennel, carrot *etc.*

c) Compound Samara (double/triple samara): The fruit is winged. It splits up at maturity into single seeded winged mericarps *e.g.* double samara of dodonaea.

d) Carcerulus: It develops from a superior bi- to multilocular ovary. It breaks up at maturity into single seeded indehiscent parts or mericarps *e.g.* garden nasturtium.

e) Regma: This is a dry 3- to many chambered fruits which develops from a syncarpous pistil. It splits away from the central axis into as many parts (cocci) as these are carpels, each part containing 1 or 2 seeds *e.g.* castor, jatropha.

Simple Succulent Fruits (Fleshy Fruits)

In these simple fruits pericarp and its associated parts become fleshy. These are of three types – berries, drupe and pome.

1) Berries or Bucca Fruits

In berries pericarp is fleshy and differentiated into epicarp, mesocarp and endocarp. The endocarp is either pulp like in grapes, guava, banana, tomato *etc.*, or papery like in dates. The fruits are single seeded or many seeded. The seeds become free from placenta and lie freely in the pulpy endocarp.

i) Superior or true berries: Developed from the superior ovary and all three layers (when floral whorl diverse below the gynoecium- flower-hypogynom) of fleshy pericarp are edible *e.g.* grapes, tomato, dates *etc.*

ii) Inferior/false/pseudo-carpic berry: Berries derived from inferior ovaries (when free floral whorl arises from top of the ovary, the ovary is called inferior and flower is called epigynous). In this case, epicarp is fused with thalamus to form the skin, also called exocarp. Remains of calyx appear as a crown at the top *e.g.* guava, banana.

Five types of special berries occur in nature:

i) Balausta: False berry in which syncarpous pistil consisting of two rows of fused carpels, one above the other. The whole fruit is covered by a hard rind made up of exocarp and part of mesocarp. Mesocarp also forms plate like infolding. The endocarp is papery and covers the individual groups of seeds *e.g.* pomegranate.

ii) Pepo: Faba berry in which the rind or exocarp does not separate from mesocarp. Similarly, seeds do not separate from the placentae which are inverted to form septa *e.g.* gourd, cucumber, melon, pumpkin *etc.* Rinds may or may not be edible. Normally endocarp, mesocarp as well as placentae (except melon) are edible.

iii) Amphisarca: Multilocular and many seeded superior berries in which outer part of pericarp or epicarp is woody, mesocarp is fibrous while endocarp and placental form the edible pulp *e.g.* bael.

iv) Hesperidium: Many chambered superior berry in which other chambers covered by its own papery endocarp. Epicarp and mesocarp fused together to form a separable skin or rind of the fruit. It is derived from a polycarpellary syncarpous superior multilocular ovary with axile placentation *e.g.* orange, lemon, pummelo.

Fig 2.8: Hesperidium (Lemon)

v) Fibrous berry: It is found in betel nut (supari). Fruit has a thin epicarp, thick fibrous mesocarp and an edible endocarp which is fleshy in moist condition but becomes hard when dry.

2. Drupe or Stone Fruit

Pericarp (forms an outer skin may or may not be edible) differentiated into three layers. Epicarp/exocarp, mesocarp and endocarp. Mesocarp is fleshy or fibrous, endocarp is stony. It encloses one or rarely more seeds (two in ber, three in borassus) *e.g.* mango, peach (only mesocarp is edible)- epicarp and endocarp is removed. Coconut (fibrous drupe)- membranous epicarp, fibrous mesocarp and stony endocarp dark with three eye spots represent the remains of the styles.

3. Pome Fruits

Fruit that develops from an inferior compound ovary. The remains of sepals and stamens occur at the free end. Pericarp is not differentiated in three parts *e.g.* epicarp, mesocarp and endocarp. Outer fleshy edible part of the fruit is thalamus. Inner pericarp is divisible into five chambers containing the seeds. The chambers represent the loculi of the inferior penta-carpellary syncarpous ovary.

Aggregate fruits (etaerio – aggregate of simple fruits borne by a simple flower): An aggregate fruit is collection of simple fruits developing from an apocarpous (free carpels) pistil of flower. Important aggregate fruits are as follow:

i) Etaerio of achenes: It is an aggregate fruit in which the individual fruitlets are achenes (dry single seeded indehiscent fruitlets in each of which the seed is attached to the dry pericarp at one point *e.g.* strawberry has brownish achenes embedded over the surface of fleshy and edible thalamus).

ii) Etaerio of samaras: *e.g.* ailanthus.

iii) Etaerio of follicles: *e.g.* larkspur, calotropis.

iv) Etaerio of berries: *e.g.* custard apple – in custard apple (sitaphal) individual berries are fused together around an elongated central axis.

They are demarcated from outer side by hexagonal areas. Edible parts are the juicy mesocarp of individual berries.

v) Etaerio of drupes: *e.g.* raspberry, blackberry.

Composite or multiple fruits: (Given at Sr. No. 3 of this chapter)

Chapter 3

Climate and Soils for Horticultural Crops

Climate

The climate of a given place is average of weather experienced in that region or area over a period of time (average over 30 years).

Various elements of climate are rainfall, temperature, humidity, air pressure, winds, clouds and sun shine *etc.*

- ☆ Good quality and yield can be achieved in an appropriate climatic condition. Fruit crop for a specific area should be selected on the basis of climatic conditions of that particular area.
- ☆ Little human efforts can control unfavourable climatic and soil conditions practically.
- ☆ Based on the climatic point of view a region can be broadly classified into tropical, subtropical and temperate zone (defined in chapter 2). However, grape can be grown in temperate and subtropical and papaya can be grown in tropical and subtropical.

Temperature

Optimum temperature for a growth of fruit plants ranges from 15 to 35°C.

- ☆ Temperate fruit crops are resistant to cold.
- ☆ Low temperature injury is of two types: Chilling injury in temperature ranges from 33 °F to 42 °F or 0.5 to 5.5°C and freezing injury occurs at temperature 32 °F or less or 0°C.

- Banana plants suffer chilling injury at 4°C.
- Pollen tube of apple germinate above 5.4°C, while it fails to germinate below 4.5°C.
- Low temperature during ripening of grape increases the malic acid (more acidic) content in the fruit.
- Litchi flowers best with day temperature below 20°C.

Influence of Low and High Temperature on Fruit Trees

Plants are to be protected against extremely low and high temperature.

Effect of Low Temperature

The trees deficient in mineral, weak, diseased, unirrigated and incompatible to rootstock are more sensitive to stress.

- In general, the capacity of a plant to retain moisture against the extracting forces of freezing is associated with its ability to survive against cold.
- Ice formation in intercellular space (angle is 109.5°) extract water (angle is 105°) from the protoplasm causes the cell die and loss of water.
- The plants which retain moisture for a longer period at a low temperature are hardy or resistant.

Black heart: Inner wood become black (often shiny black), while the cambium and the bark remain alive. Nursery plants are susceptible to such type of injury.

Crotch injury: Killing of tissue in the fork or crotches of both small and large branches.

Crown and collar injury: Frost damage the stem at ground level, bark of young trees is killed, cracked open and inner sap carrying tissue is ruptured through freezing. It extends towards roots.

Remedies: Remove the dead bark and paint the wound to prevent the entry of the fungus/water.

Follow the bridge grafting using hardy rootstocks.

Splitting of the bark: It is caused by severe frost injury and extend to the entire trunk or branches under extreme conditions.

Killing back of shoot and young branches: Tropical and subtropical trees may be damaged by frost in early winter or early spring. Bark along the cambium layer is discoloured or killed *e.g.* mango flowers may be killed.

Citrus and papaya fruits are injured.

Frost damage: Evergreen trees suffer from the frost injury. Young plant and the shoots may damage terminally. The trees which bear fruits during winter suffer more.

- Frost affected fruit shows dry patches and have tasteless juice.

Protection against Cold

Cultural Practices

- Frost tolerant varieties.

  Mango: Langra, Bombay Green, Safeda, Fajri Kalan and Malda.

  Frost tolerance citrus in descending order: Mandarin, khatta, sweet orange, pummelo, galgal, sweet lime, kagzi lime.
- Tender fruit plants should be planted in the inner side of the orchard.
- Small plants and nursery plants need to be protected with hedge.
- Adequate nutrient application, plant protection measure and timely training, pruning and irrigation - as the healthy plants are tolerant to adverse weather/low temperature.
- Cover young plants with thatches made of farm waste.
- Irrigate the orchard during winter, it raises the temperature by 2 to 3°C.
- Plant windbreaks on the North-Western sides of the orchard, which provide protection from cold waves during winter season.

Mechanical Management

- Heaters: Heaters are frequently used in some foreign countries to heat the citrus orchards. However, under Indian conditions, it will be very expensive.
- Covering/wrapping: Wrap exposed portion of tree with old gunny cloth to avoid damage to bark by cold injury.
- Heating by smoking: Dry leaves can be burnt as smoke raise the temperature by few degrees and save fruit trees from frost (burning of crop residue has been banned by Govt.).
- Steam heating: It is expensive so not advocated.
- Shelter: Cover the plants with kans (*Saccharum spontaneum*), maize, jawar stalks. Tender nursery plants can be grown in green house.

Effect of High Temperature

- Rapid transpiration desiccates the leaves and twigs.
- Excessive heat during growing season results in burning of leaves and fruits.
- During the day, branches, leaves, fruits and stem portion exposed to sun become hotter than the surrounding atmosphere.
- Heat injury become more harmful when accompanied with low humidity and strong winds due to rapid transpiration/desiccation.

Sun burn: Excessive heat during the growing season often results in the burning of leaves and fruits. During the day, branches, leaves, fruits and stem portions exposed to sun become hotter than the atmosphere, resulting in this sort of injury. Heat is generally more harmful when it is accompanied by low humidity and strong winds. These conditions cause rapid transpiration and desiccate the leaves and twigs.

Sun scalds - Sun light is a good source of energy and colour development while strong sunlight towards south western facing side of the tree is called sun scald. It is more common in citrus and bael.

- **Low headed trees**: The plants trained to low headed frame work escape the ill effect of the hot sun better than the high headed trees. It is, therefore, desirable that the fruit trees should be trained to develop low heads, especially in the hot and arid tracts.
- The fruit trees, especially young plants need attention and adequate protection against hot summer and winter frost. The fruit plants like mango, citrus, litchi are damaged more by high summer temperature.

Remedies against High Temperature

Cover with rapping material: Any ordinary or cheap paper can be used to wrap around the trunk.

White washing: This is most practiced method for saving trunk of the fruit trees from sun injury. The bare trunk portion should be white washed during March, it may again be repeated if it gets washed by rains. The effective white wash material can be prepared by adding unhydrated lime (25 kg) and copper sulphate (2 kg) in 450 litre of water as copper sulphate is effective against fungal diseases.

Providing shelters/shedding: Young tender fruits plants can be protected by erecting thatches made of paddy straw.

Wind break: The windbreak should be planted before fruit tree plantation to protect the plants from high wind velocity.

Frequent irrigation: Frequency of irrigation should be increased during summer to meet out evaporation losses.

Atmospheric Humidity

- High humidity favour the insect-pest, disease and adversely affect colour and quality of the fruits.
- High humidity during flowering and fruiting will promote diseases like mildew (mango and grapes) if atmospheric humidity is increased by rains, it causes fruit crack in pomegranate and lemon; and fruit spilling in fig.
- Incidence of diseases like gummosis and citrus canker is also favoured by high humidity.

Rainfall

In arid and semi-arid zone the production of fruits mainly depends on rainfall.

- ☆ Heavy rainfall adversely affects the blossoms and fruits.
- ☆ Heavy rains during flowering cause washing off pollen, hinder insect pollination, injure pollen, dilute stigmatic fluid and promote incidence of pest and diseases.
- ☆ Rainfall before harvesting also decrease the fruit quality and TSS of fruits.

Hails: It is also limiting factors in fruit growing.

- ☆ Remedy: Grow small/tender crops under protected cultivation.

Winds

- ☆ High wind velocity cause uprooting of plants.
- ☆ Dropping fruits and flower.
- ☆ Prevent pollinator.

Remedies: Windbreak and shelterbelt.

Wind break: It depends on the height and density/intensity of the trees, wind break is effective for a distance of 3-4 times of its height.

- ☆ Protect orchard from high wind velocity.
- ☆ Reduce velocity of wind.
- ☆ Check evaporation losses.
- ☆ Prevent from cold/hot wind.
- ☆ Reduce frost damage.

  e.g. Polyalthia longifolia, Casuarina equisetifolia, Erythrina indica, Eucalyptus globules, Syzygium sp., *Grevillea robusta, Mangifera indica, Dalbergia sissoo, Bambusa* sp., *Putranjiva roxburghii, Averrhoa carambola etc.*

Fencing: To protect the trees or orchard from the frequent visit of wild and stray animal and prevent stealing of fruits and other properties. This may be made from-

- ☆ Erratic mud wall/brick wall.
- ☆ Barbed wire fencing.
- ☆ Live hedge (bio-fencing): Growing of plants having following characteristics.
  - ❖ Quick growing
  - ❖ Easy to raise by seed or cutting
  - ❖ Drought resistant
  - ❖ Dense foliage
  - ❖ Thorny

❖ Develop thick and compact growth of canopy

e.g. Carissa carandas, Casuarina equisetifolia, Inga dulcis, Prosopis juliflora, Parkinsonia aculeata, Duranta alba, Bauhinia sp, Polyalthia longifolia etc.

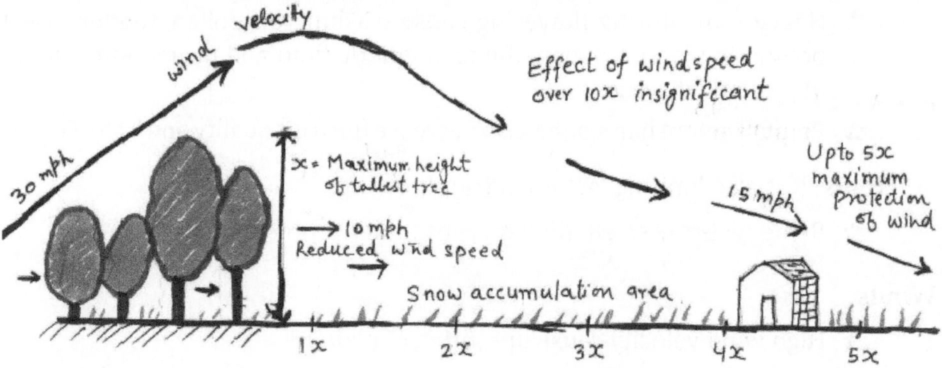

Fig 3.1: Effects of a Windbreak on Wind Speed

In addition to above the growers should remain alert and vigilant, keep in touch with the weather forecasts or frost warning by radio, TV or weather forecasting message service. The grower should install the thermometer in their orchards for knowing fall in temperature. When the temperature falls around 0 °C and the atmosphere is still and sky is clear, the frost is sure to occur. Farmers can also install the Mobile App of IMD or CCS HAU, Hisar for weather forecasting.

Flood Stress

The reduction of oxygen below optimal levels is called hypoxia. It is a most common in wet soils and occurs during short term flooding when the roots are submerged and shoots remain in the atmosphere. Hypoxia will occur in roots near the surface of long-term flood water. The complete lack of oxygen is called as anoxia it occurs in soils that experience long term flooding or plants remain completely submerged in water. Long term flooding shifts the microbial flora.

Drought

Scarcity or shortage of water in the root zone of the plant which check the plant growth. The losses due to drought depend on the duration and intensity of the drought.

The severity of drought depends on factors like distribution of rainfall, evaporative demands and moisture storing capacity of soils. Drought classified into two broad categories: i) Soil drought ii) Atmospheric drought.

Main mechanisms in reduction of crop yield by soil water deficit: (i) reduced absorption of photosynthetically active radiation in canopy, (ii) decreased radiation-use efficiency and (iii) reduced harvest index.

Plant Response against Drought

Morphological symptoms: Wilting of leaves and twigs, reduced growth and yield.

- ☆ Fruit cracking decreased yield.

Anatomical changes: Reduced cell size, cell division, cell enlargement and differentiation.

- ☆ Closure of stomata - limits CO_2 uptake by leaves.
- ☆ Damage the mesophyll cell.

Phenological changes: Flower abortion.

- ☆ Fruit drop

Physiological Responses

Water relations: Reduced water availability and uptake of nutrients.

- ☆ Decrease in water potential.
- ☆ Reduced transpiration.

Biochemical changes: Increase in glucose and fructose concentration with decrease in starch content.

- ☆ Decrease in protein and increase in proline concentration.

Drought Resistance Mechanisms

The capacity of a plant to grow normally in a dry habitat and giving maximum yield. In true sense, drought resistance is the capacity of plants to tolerate drought and to recover rapidly after the onset of permanent wilting with minimum damage to plant itself. Drought is a period or condition of unusual dry weather within a geographic area where there is a lack of precipitation. It is governed by various factors, the most prominent being extremes of temperature.

Drought tolerance is the ability of a plant to grow, flower and produce economic returns under deficient water conditions. Drought stress affects the water relations of plants at cellular, tissue and organ levels.

1. Morphological Mechanisms

Plant drought tolerance involves changes at whole-plant, tissue, physiological and molecular levels. Manifestation of a single or a combination of inherent changes determines the ability of the plant to sustain itself under limited moisture supply.

i. **Escape:** Drought escape occurs when phenological development is successfully matched with periods of soil moisture availability, where the growing season is shorter when terminal drought stress predominates. Shortened the life cycle or growing season, allowing plants to reproduce before the environment becomes dry and shedding of the leaf. Developing short-duration varieties has been an effective strategy for minimizing yield loss from terminal drought, as early maturity helps the crop to avoid the period of stress.

ii. **Avoidance:** Reduce water loss from plants by stomatal control of transpiration. Maintain water uptake through an extensive and prolific root system (deep and thick root system). Waxy layer on leaves helps with maintenance of high tissue water potential, and leads to drought tolerance.

iii. **Phenotypic flexibility:** Plant growth is greatly affected by water deficit. Plants generally limit the number and area of leaves in response to drought stress just to cut down the water budget at the cost of yield loss. Since roots are the only source to acquire water from soil, hence; the root growth, its density, proliferation and size are key responses of plants to drought stress.

2. **Physiological Mechanisms**

i. **Cell and tissue water conservation:**

Osmotic adjustment - Decrease osmotic potential and increases the gradient for water influx and maintenance of turgor.

High tissue water - Essential for maintaining extended periods of physiological activity during drought.

Osmotic adjustment, abscisic acid and induction of dehydrins may confer tolerance against drought injuries by maintaining high tissue water.

ii. **Antioxidant defense:**

Enzymatic components: Superoxide dismutase, catalase, peroxidase, ascorbate peroxidase and glutathione reductase.

Non-enzymatic components: Cystein, reduced glutathione and ascorbic acid.

iii. **Cell membrane stability:** Maintenance of integrity and stability of membranes under water stress is a major component of drought tolerance in plants.

- ☆ Membrane disruption
- ☆ Decrease in cellular volume
  - Protein denaturation
  - Membrane fusion
- ☆ K nutrient- cell membrane stability

iv. **Plant growth regulators:** Under drought, endogenous contents of auxins, gibberellins and cytokinin usually decrease, while those of abscisic acid and ethylene increase. Nevertheless, phytohormones play vital roles in drought tolerance of plants.

- ☆ Calcium and auxin participate in signaling mechanisms of drought-induced proline accumulation.
- ☆ Increase in abscisic acid and decline in cytokinin levels favour stomatal closure and limit water loss through transpiration under water stress.
- ☆ Abscisic acid alters the relative growth rates of various plant parts such as increase in the root-to-shoot dry weight ratio, inhibition of leaf area development and production of prolific and deeper roots.

v. **Compatible solutes and osmotic adjustment:** Under water deficit and as a result of solute accumulation, the osmotic potential of the cell is lowered, which attracts water into the cell and helps with the maintenance of turgor.

3. Managing Drought Stress

Drought stress effects can be managed by selection of resistant crop/variety/rootstock and some management practices.

Choice of Crop Variety/Rootstock

Choice of arid zone crops, which can survive easily under water stress condition *e.g.* ber, aonla, karonda, phalsa, date palm, bael. Crops can be selected on the basis of the following criteria:

- ☆ Crops having deep root system: Mango, ber, wood apple, bael.
- ☆ Less in number and small sized leaves: Aonla, tamarind, kair, karonda.
- ☆ Arrangement of aerial shoot (dense): Mango, bael, wood apple.
- ☆ Leaves densely pubescence: Phalsa and fig.
- ☆ Have shining surface due to presence of thicker cuticle: Sapota, cashewnut, pomegranate.
- ☆ Radiation reflection by waxy coated leaves: Ber, bael, wood apple, fig.
- ☆ Decreased stomatal density and frequency: Ber, cashew nut.
- ☆ Sunken deep stomata: Ber.
- ☆ Reduction in cell and stomata size: Ber.
- ☆ Low osmotic and water potential: Ber, pomegranate.
- ☆ Secretion of latex in leaves: Fig, karonda.
- ☆ High degree of phenol, tannins and oil content: Mango, cashewnut, karonda.

Use drought tolerant rootstocks:

Grapes – Dogridge, Salt Creek, 1616, 1613.

Citrus - Rangpur Lime, Rough lemon.

Most of the indigenous plant rootstocks are tolerant to drought and salinity.

Management Practices

Planting: 2-3 cm thick layer of bentonite clay or pond silt is placed in bottom and side of planting pit reduce water losses through infiltration.

To develop hardy trees, *in situ* raising of plants should be preferred.

Weeding: Weeds are severe competitors for water and nutrients. It takes about 80 gallons of water to grow one pound of weeds.

Ways to control weeds: Mechanical, chemical/herbicides, mowing and combinations of these practices.

Use of shed-nets in nursery: Keeping nursery plants under shed-nets and irrigating them through micro-jets facilitates to increase survival percentage and reduces irrigation needs.

Micro-irrigation: Drip irrigation has been used extensively by fruit growers and it improves water use efficiency (80-95 per cent) and total water use.

Drip is a new way to irrigate crops which prevent crops from water stress. However, drip does not reduce the water requirements of plant.

Sprinkler irrigation: Sprinkler irrigation offers the ability to irrigate uneven terrain effectively. However, the energy required to pressurize the water can become expensive.

The new low energy precision application (LEPA – irrigation applied on the soil surface at very low pressure which minimize the evaporation and wind drift losses) systems are getting popularity.

These systems have very high initial cost but have tremendous efficiency.

Few growers can afford improved irrigation systems during drought stress period.

Mulching: Cover the soil surface by plastic, crop residue and growing cover crops or inter-culturing in the orchards. These methods also check soil erosion and runoff of rainwater and evaporation losses.

i) **Plastic mulch:** Soil moisture can be conserved through mulching with black polythene sheet of 100 micron or locally available mulches in the orchards. It reduces evaporation losses of water and check weed population by 100 per cent. These methods also check soil erosion and runoff of rainwater.

ii) **Organic mulch:** Growing cover crop/intercrop or covering soil with crop residues is called organic mulching. Sometimes, plants are covered with mulches preferably grass mulch (10-12 kg/basin). This helps in conserving soil moisture by reducing evaporation losses of water and reducing the weed population by 60 per cent. This also reduces the cost of cultivation.

Fig 3.2: Plastic Mulching

Fig 3.3: Organic Mulching

Anti-transpirants: Anti-transpirants are the materials or chemicals which decrease the water loss from plant leaves by reducing the size and number of stomata. Nearly 99 per cent of the water absorbed by the plant is lost in transpiration.

Different Anti-transpirants

1. Metabolic inhibitor like phenyl mercuric acetate, succinic acids.
2. Growth retardant such as ABA, cycocel.
3. Herbicides (Atrazine), fungicides (phenyl mercuric acetate).
4. Salicylic acid.

5. Colourless plastics, silicon oil, wax or plastic.
6. White reflecting materials (*e.g.* kaolin) emulsions or white wash.

Water harvesting: Collection of rain water for further use or recharge ground water. This method consists of in situ water harvesting through opening of small trenches and contour bunding.

A catchment is an area where water is collected by the natural landscape. In a catchment, all rain and run-off water eventually flow to a creek, river, dam, lake, ocean, or into a groundwater system. The water from catchment area or stagnated in low lying area of farm can be used to irrigate the crop as life-saving irrigation.

Pruning practices/canopy control: Pruning reduces leaf area in fruit crops and decrease transpiration loss of water hence conserve water under stress.

Summer pruning can also be used in apple to conserve moisture and severe pruning of peach and pear has been found effective to reduce transpiration.

Nutritional management: Potassium and magnesium enhances the water uptake as well as the water relations in the plant tissue by osmo-regulation process during water deficit conditions.

Magnesium is important component of chlorophyll; its content and uptake are drastically reduced due to the water stress effect.

Urea sprays - Enhance drought tolerance by decreasing osmotic potential.

Foliar spraying of micronutrients *i.e.* boron and zinc, which becomes unavailable in water stress conditions.

Calcium treatment

External Ca^{++} treatment -increase activity of catalase (CAT), superoxidase dismutase (SOD) and peroxidase (POD). Ca^{++} treatment improves adaptation to drought by mitigating oxidation stress.

Plant growth regulators

Triazoles: A group of plant growth retardants which improves plant survival under low soil moisture conditions. Triazol compounds are Paclobutrazol, Triapenthenol, Uniconazol and BAS -111.

Mode of action

Interfere with biosynthesis of gibberellins by inhibiting the oxidation of Kaurene to Kaurenoic acid by inhibiting the activities of cytochrome P-450 dependent oxygenases.

Increases the level of ABA, Triacontanol (TRIA) and crystalline substances.

Stimulate photosynthesis which increase water uptake and dry weight/biomass.

Application of Paclobutrazol (10 mg/l) is used to avoid moisture stress in mango and exogenous application of jasmonic acid induced drought tolerance by increasing the betaine level in pear.

Molecular and Functional Genomics Approaches

Transgenic expression of some of the stress-regulated genes results in increased tolerance to drought and other stresses. These transgenic approaches are currently mainstream method to bioengineer drought tolerance in crop plants

Fig 3.4: Water Stagnation-Collar Rot in Papaya

Fig 3.5: Low Temperature - Dying of Aonla Plant

Fig 3.5 (a): Initial Symptom of Low Temerature Injury in Aonla

Fig 3.6: Salinity Effect on Mango Seedling

Fig 3.7: High Temperature Injury in Ber and Guava Seedlings

Fig 3.4 (a): Collar Rot in Papaya

- Drought tolerance is a genetically controlled phenomenon, many quantitative trait loci for membrane stability and other functionally related phenomena genes have been characterized using bioinformatics tools.
- The combination of traditional breeding, molecular breeding, marker-assisted selection and genetic engineering may allow more rapid way to improve abiotic stress tolerance in fruit crops.

iii. Induction of Drought Resistance

a. **Seed priming:** Seed priming is soaking of seed to initiate the germination process under water deficient conditions to overcome short-term moisture deficiency for germination of seed.

b. **Use of plant growth regulators:** Application of growth regulators synthetic as well as natural is helpful to improve the growth of the plant under abiotic stress conditions. Drought stress inhibits the increase in length and fresh weight of hypocotyl, whereas gibberellic acid reversed this effect.

c. **Silicon:** Silicon plays an important role in growth, nutrition and mechanical strength under stress conditions. It is most abundant and easily available element in the soil.

Soil

Porous, aerated and deep soil should be preferred for fruit cultivation. The ideal orchard soil should be at least 1.8 m deep having uniform texture, well drained, non-saline and fertile without rocky substrate. Most of the fruit plants like slightly acidic to neutral soil reaction (pH 6-7).

- For mechanical support to plant.
- Provides air, water and nutrients to the plants.

Common soil management practices are:

- **Tillage**: The objective of cultivation is to loosen and pulverize the soil so that the roots of fruit plants may penetrate in the soil easily. It facilitates circulation of air and absorption of water. Tillage removes the weeds and help in mixing manures and fertilizers.
- **Sod culture**: Plants are allowed to grow in permanent grasses without tillage and addition of litter. It checks soil erosion and damage to fallen fruits. But water and nutrients are taken by weeds. Hazards of fire and attack of more insect pest is also more.
- **Mulching**: It is a soil management technique in which hay, straw, or cut grasses are placed on the soil between the tree rows to conserve soil moisture and checking evaporation and germination of weeds. Use of cut over grasses as a mulch is called sod mulch.
- **Clean cultivation**: Keeping the orchard free from grasses or any other vegetation.
- **Clean basin management**: This practice is carried out manually by keeping the basin free from all weeds.

A) Selection of site : Fruit crops can be grown on wide range of soils, but for better production soil should have following characteristics:

- ☆ Well drained.
- ☆ Free from hard pan for proper root penetration.

i) Soil texture and depth

- ☆ Sandy loam/loam soil.
- ☆ Soil depth should be more than 1.8 m.
- ☆ It should be porous.

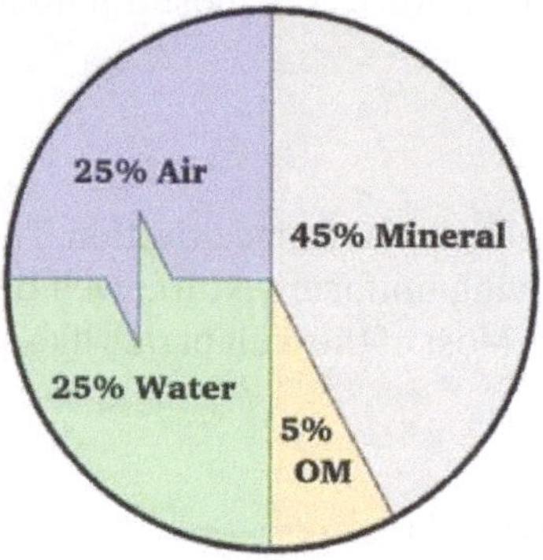

Fig 3.8: Soil Composition

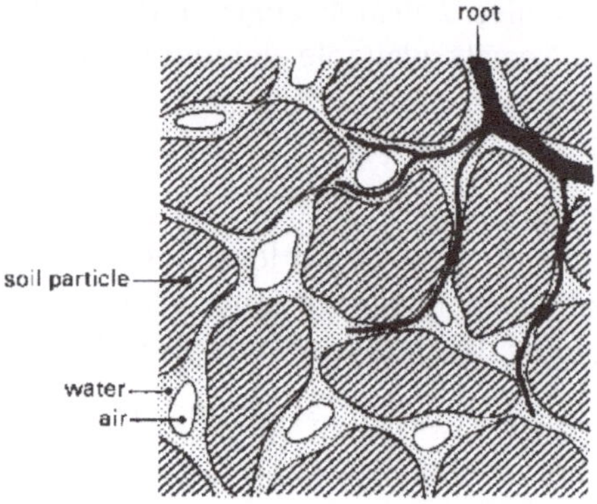

Fig 3.9: Soil Porosity

ii) Poor drainage of soil

- ☆ High water table
- ☆ Compact soil
- ☆ High clay content
- ☆ Hard pan

iii) Soil pH

It is used to measure the acid or alkali level in the material or soils. Most suitable soil pH for fruit crops is 6.5 to 7.5.

Suitable pH for availability of nutrients

pH less than 5: Al and Mn toxicity often occurs in roots and also cause deficiency of Ca, Mg and P.

Reclamation: Apply lime stone or dolomite.

Avoid N in NH_4 form (ammonium sulphate).

Citrus: pH more than 7.5 suffer from deficiency of micronutrients (Fe, Mn, Cu and Zn).

Reclamation: Foliar application of nutrients.

High pH: Lower availability of plant nutrients *e.g.* Mo.

Reclamation: Apply gypsum (replace Ca with Na).

$Na_2CO_3 + CaSO_4 \rightarrow CaSO_3 + Na_2SO_4$ (leachable).

iv) Electrical conductivity (EC)

It is used to measure the ability of material to carry electric current or capability of water to pass electric current (dS/cm or mmhos/cm).

Depth for samples from ground level is given below:

1. 0-15 cm
2. 15-30 cm
3. 30-60 cm
4. 60-90 cm
5. 90-120 cm
6. 120-150 cm
7. 150-200 cm

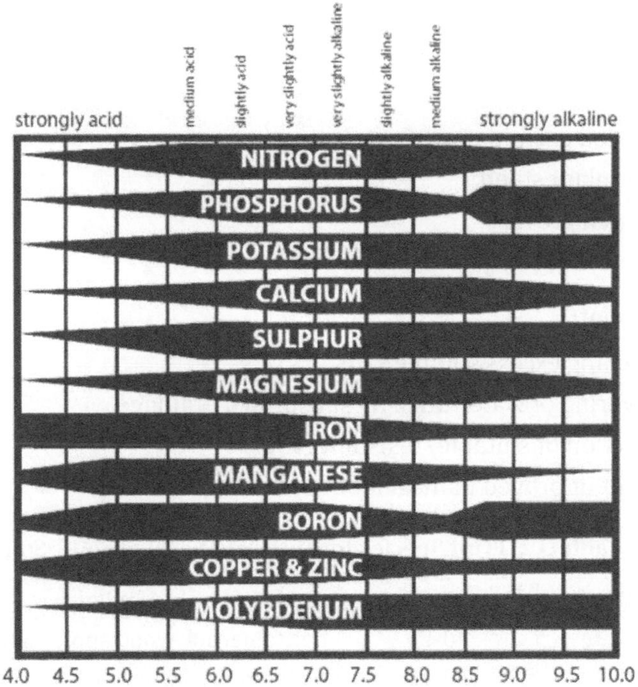

Fig 3.10: Nutrient Availability vs. Soil pH

*(**Source:** Reitsona et al. 2011)*

Precautions

– Take separate sample for stone or hard pan if any along with depth and thickness of pan.

– Keep each sample in separate bag.

– Take at least 5 samples per acre.

Table 3.1: Appropriate Physical and Chemical Quality of Soil for Orchard

Quality	*Citrus Group*	*Other Crops*
EC mmhos/cm (1:2)	0.5	1.0
Calcium carbonate (per cent)	5	10
Stone of calcium carbonate (per cent)	10	20
pH	8.5	Max 8.5 (EC less than 0.5)

v) Salt content

Arid zone soils are largely saline and underground water is also saline.

Major salts: Chloride and sulphate of N, Ca and Mg.

- ☆ Osmotic effect reduces the availability of nutrients and water.

Excess salt:

- ☆ Delayed germination
- ☆ Poor plant stand
- ☆ Stunted growth
- ☆ Reduced yield

Management:

- ☆ Leaching excess soluble salts
- ☆ Lowering of water table by subsurface drainage
- ☆ Selection of suitable/resistant crops
- ☆ Adopt improved management practices

Table 3.2: Fruit Species for Different Level of Sodic Soil

Tolerance of Sodicity	*ESP*	*Fruit Crops*
High	40-50	Ber, tamarind, wood apple, date palm
Medium	30-40	Aonla, karonda, fig, phalsa, pomegranate
Low	20-30	Guava, bael, lemon

Table 3.3: Fruit Species for Different Saline Conditions

Salinity Level	*EC dSm^{-1}*	*Fruit Crops*
Plant species for slightly saline conditions	1.0-2.0	Bael, fig, phalsa, jamun
For moderate saline	2-4	Aonla, wood apple
For highly saline	4-10	Khirni, date palm, ber
Very highly saline	10-25	Ker, khejri
Species for saline	25-38	Pilu

Management of Salinity

- Maintenance of the soil organic matter.
- To enhance physical, chemical and biological properties of the soils.
- Add livestock manure, green cover, crop residue, mulch and weeding.

☆ Improve fertility: Balanced fertilizer application.
- Excessive use may damage the fruit quality and reduce the yield.

☆ Green manuring: Cluster bean, dhaincha, sunhemp.
- Neem leaf, mahua, khajri, subabool.
- It helps in reclamation of salt and manage the root knot nematode.

☆ Sod culture: To control soil erosion in hilly areas or in high soil erosion areas.

Grow grasses between the trees to avoid soil erosion.

☆ Inter-culture: It depends on the nature of main crop.
- Suitable up to 3-4 per year.

☆ Cover crop: Cowpea, moong, soyabean.
- Check erosion
- Improve soil physical conditions
- Add organic matter
- Maintain moisture
- Maintain soil temperature

☆ Clean cultivation:
- Keep orchard free from grasses
- Control weeds
- Improve moisture retention
- Improve nutrient and water availability
- Improve soil aeration

- ☆ Water table: Should be more than 2 m.
  - *e.g.* Date palm can tolerate water stagnation.
  - Papaya and guava are sensitive to water stagnation.
- ☆ Mulching:
  - Organic
  - Inorganic (black polythene)
- ☆ Tillage: Aeration
- ☆ Minimum tillage: *In-situ* plant residue decomposition

Chapter 4

Propagation Methods

Nursery

A specially designed/managed site used to produce seedlings under favourable conditions till they are not ready for transplanting.

Nursery Site Selection

- The site should be accessible round the year for customers and transport, preferably on metalled road.
- Reliable supply of water, ideally being near a river or ponds.
- Good soils and other planting materials, sand, FYM, pots and other inputs should be available easily.
- The site should be protected from strong winds and livestock.
- It should receive good sunlight.
- Soil should be well drained, fertile, level and free stones.

Propagation Methods

Multiplication or reproduction of plants using various techniques or methods or procedure is called propagation. Propagation is of two types:

1. Sexual propagation
2. Asexual propagation

Sexual Propagation

Seed are formed as a result of fertilization of male and female gametes so it is called sexual propagation *e.g.* papaya, phalsa, arecanut, coconut, cocao and coffee.

Seed

It is ripened ovule developed after fertilization of male and female gametes. It has three important parts embryo (part to initiate root and shoot), endosperm (having stored food to be supplied during germination) and seed coat (covering or protection of seed). Seed germinate into a plant during or when provided favourable growing conditions.

Fig 4.1: Karonda Seedlings

Merits of Sexual Propagation

- Plants produced from seeds are harder and have long life span.
- Storage for longer time and transportation of seed as a planting material is very easy.
- Plants which cannot be produced easily through asexual method *e.g.* coconut, papaya, phalsa *etc.,* can only be multiplied by seed.
- Seedlings for rootstocks are obtained by growing seeds for budding and grafting.
- Seedling plants are resistant to insect-pests and diseases.
- It is the easiest and least expensive method of plant multiplication.
- In case of polyembryonic varieties (emergence of more than one seedling from one seed) true to type seedlings can be prepared by seeds.
- Maintain or create genetic diversity is major challenge which can be achieved by sexual propagation.

- ☆ Hybridization process for release of new varieties can be achieved by sexual propagation only.

Demerits of Sexual Propagation

- ☆ Seedlings have more gestation period so take more time to bear fruits.
- ☆ This method is not effective to improve the quality of existing plants.
- ☆ Sexually propagated plants are vigorous, thus the cost of operations/ application of manuring, pruning, spraying and harvesting increases.
- ☆ Most of the fruit crops are cross pollinated so the genetic purity cannot be maintained in sexual propagation.
- ☆ Sex identification in seedling of dioecious plant is not possible.
- ☆ Additive effect of rootstock on scion can be achieved by vegetative propagation.
- ☆ It is not possible for crops which are not producing seeds or viable seeds to propagate through sexual method *e.g.* banana, fig, pineapple, strawberry, hibiscus, jasmine, bougainvillea *etc.*

Asexual or Vegetative Propagation

Different parts of plants like root, stem, leaf and their modified forms are used to multiply the plant is called asexual propagation. To produce true to type planting material most of the horticultural crops propagated asexually.

Merits of Vegetative Propagation

- ☆ It produces true to type and uniform plants.
- ☆ Seedless plants *e.g.* pineapple, banana, seedless grape varieties and seedless guava can only be produced vegetatively.
- ☆ Precocity of bearing.
- ☆ Manageable size and uniform fruits.
- ☆ Grafting on resistant rootstock *e.g.* Mandarin is budded on Rangpur lime rootstock to avoid gummosis.
- ☆ It is helpful to repair damaged portion of plant through bridge grafting.
- ☆ Multiple varieties can be propagated on single/same plant.
- ☆ HDP can be recommended due to small canopy and restricted growth of scion/rootstock.
- ☆ This method is used for rapid multiplication by using plant tissue culture.

Demerits of Asexual Propagation

- ☆ Vegetatively propagated plants sensitive to adverse conditions of soil, climate, insect-pests and diseases *etc.*
- ☆ Less chances of heterogeneity in planting material.

- Multiplication of plants like arecanut, papaya, coconut *etc.* is expensive and difficult in asexual propagation.
- Vegetative propagated plants are generally less vigorous and require special skill and time for multiplication; however, seedlings are vigorous and long lived.
- Improvement of crop through hybridization is not possible by asexual method of propagation. New variety can not be developed by this method.

Propagation by Cutting

Cutting

Vegetative plant part(s) that has ability to develop into a new plant without specialized skills. It is quick method of multiplication which requires less skill and production cost. In this method more number of true to type plants can be produced using different parts from mother plants. Based on the part of plant, these are described as under:

- Stem cutting
- Root cutting
- Leaf cutting

Stem Cutting

Propagation from stem cuttings is probably the most important propagation technique in use throughout nursery industry. It is used for multiplication of ornamental trees, shrubs, ground cover plants, herbaceous perennials, alpines and indoor plants. On the basis of age and strength of cutting these are descried as:

i. Hardwood Cutting
ii. Semi-hardwood Cutting
iii. Softwood Cutting
iv. Herbaceous Cutting

i) Hardwood Cutting

Woody cuttings of about one year old are taken from deciduous plants for multiplication of plants. These cutting are mostly taken during dormant season. These should be of pencil thickness with a length of 10-15 cm and planted in a rooting medium. These may be treated with rooting hormone before planting for faster rooting *e.g.* grapes, pomegranate, fig, rose, jasmine, bougainvillea, hibiscus *etc.*

Procedure

- Collect pencil thickness cutting from healthy plant. Length of cutting should be about 10–15 cm with at least 4-5 buds/nodes. To avoid transpiration losses, remove leaves and thornes.
- Give a slanting cut at lower end and straight at upper end. It will help to recognize the lower and upper portion of the cutting.

- ☆ At the same time the lower slanting cut will be helpful in increasing the area of contact with rooting medium for absorption of more nutrients and water for early rooting and sprouting.
- ☆ Upper straight cut can be pasted with wax to avoid/reduce transpiration losses. Root and shoots are induced by secretion hormones near the cut portion of the cutting.
- ☆ Cuttings are to be planted slant wise in a growing medium like beds and polybags. Cambium tissue is helpful in formation of callus for rooting and sprouting from the cutting. The season of planting cutting for evergreen plants is monsoon (July to September) and deciduous plants is November to February. For more success and better results cuttings can be planted in polyhouse or green house.

ii) Semi-hardwood Cutting

The bark of semi-hard wood cutting is slightly greenish to brownish. These cutting are taken from the current season woody growth and age of the cutting varies from 4 to 9 months. Most of the ornamental plants are propagated by semi hard wood cutting *e.g. Diffenbachia*, nerium, aralia, cestrum, croton, acalypha, russelia *etc.*

Procedure

- ☆ The semi hard wood cutting is taken from the current season woody growth.
- ☆ Cuttings should be of 7.5 to 15 cm long with 4-5 buds.
- ☆ Thickness of cutting is pencil thickness.
- ☆ Large leaves are removed and other leaves are retained for preparation of food by photosynthesis.
- ☆ Give a slanting cut at lower end and straight at upper end. It will help to recognize the lower and upper portion of the cutting.
- ☆ At the same time the lower slanting cut will be helpful in increasing the area of contact of cambium with rooting medium for absorption of more nutrients and water for early rooting and sprouting.
- ☆ Upper straight cut can be pasted with wax to avoid/reduce transpiration losses and check infection. Root and shoots are induced by secretion hormones near the cut portion of the cutting. Upper straight cut also reduces the area for exposure to open environment and also reduce the transpiration losses.
- ☆ Cuttings are to be planted slant wise in a growing medium like beds and polybags. Cambium tissue is helpful in formation of callus for rooting and sprouting from the cutting. The season of planting cutting for evergreen plants is monsoon (July to September). These cuttings are planted in mist chamber for early rooting. For more success and better results cuttings

can be planted in polyhouse or green house. Dipping in rooting hormone *i.e.* IBA @ 5000 ppm induce early rooting.

iii) Softwood Cutting

Softwood cuttings are taken from herbaceous or succulent plants. The age of softwood cuttings varies from 2 to 3 months *e.g.* duranta, tea, coleus, alternanthera, clerodendrum *etc.*

Procedure

Select a softwood cutting of two to three months old having 10-12 cm length. The leaves present on the cutting are kept for photosynthesis because these cuttings don't have reserve food material. Cutting are to be collected during early morning and kept in most cloth to avoid transpiration losses. Yamuna sand or coarse sand medium is the best for rooting in such cuttings.

iv) Herbaceous Cutting

Herbaceous cutting are very soft and tender shoots of about 1 to 2 months old taken from herbaceous plants *e.g.* carnation, chrysanthemum, petunia, marigold, pilea, dahlia, iresine *etc.*

Procedure

Soft tender leafy shoots of about 8 to 12 cm length are taken from healthy plants while taking for cutting the basal leaves are removed and upper leaves are retained for photosynthesis and supply of food material to the cutting for rooting and sprouting.

Cutting should be planted in a rooting medium (sandy loam) immediately after detachment from the plant to avoid desiccation from the cut end. Growth hormone like auxin can be applied for early rooting in the cutting.

Leaf Cutting

The thick and fleshy leaves are used for propagation of plants such leaves have buds on the margin. *e.g.* bryophyllum or in the veins begonia rex. Full leaf or leaf blade having buds are kept in rooting medium under favourable conditions.

Propagating plant using leaf blade along with petiole and small piece of wood *e.g.* black raspberry, snake plant (sansevieria), bryophyllum and rhododendron.

Plant Propagation by Layering

Layering

In this method of propagation roots are produced on the plant part attached to the mother plant which is covered with rooting medium or soil. After emergence of the roots this shoot is detached from the mother plant and grown as a separate plant on its own roots. These rooted stems are called layering.

Types of Layering

(i) Simple layering
(ii) Serpentine or compound layering
(iii) Trench layering/Etiolation layering
(iv) Stooling or mound layering
(v) Air layering

i) Simple Layering

In this layering tongue like cut is given on the stem or branch which is near to the soil surface after that branch is bent towards the soil and covered with soil (about 3 to 4 inch) keeping the terminal portion of the cutting above the ground. The underground portion of the shoots or layered branch roots than this layered portion is detached from the mother plant and ready for transplanting in field or nursery *e.g.* jasmine, clerodendrum, ixora, pyrostegia *etc.*

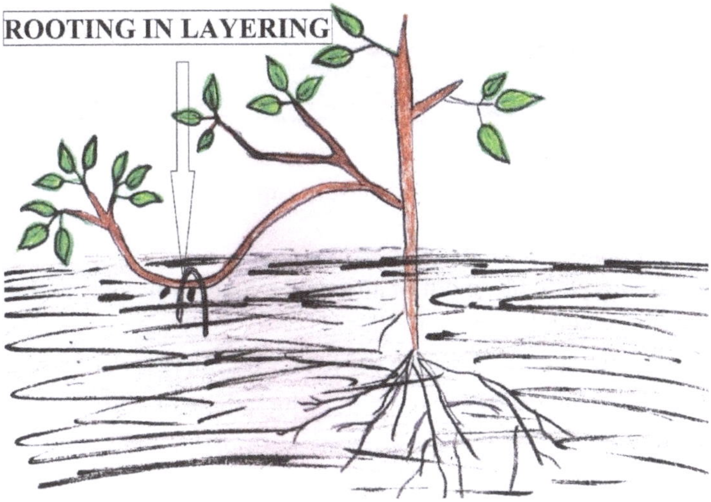

Fig 4.2: Simple Layering

Procedure

- ☆ Select one-year old shoot from mother plant near the ground level which should be flexible and branched.
- ☆ Remove leaves from the shoot retaining on the tip portion. Retained leaves will help to produce photosynthates for early rooting.
- ☆ Give a tongue cut of about 2 to 3 cm on the lower portion of the shoot.
- ☆ Bent down the shoot and cover the ground attached portion with soil retaining about 15 to 30 cm terminal end.

- ☆ Keep some weight on the buried portion of the cutting to keep in the same position or undisturbed. After 3 to 4 weeks the shoot starts sprouting into new buds is an indication of start of rooting. After that separate it and transplant in nursery or orchard.

ii) Compound or Serpentine Layering

In case of serpentine or compound layering the shoot is buried at more than 2 spots. The branch must be longer and flexible to layer at more or several places *e.g.* jasmine, bougainvillea, muscadine grape, clematis and wisteria.

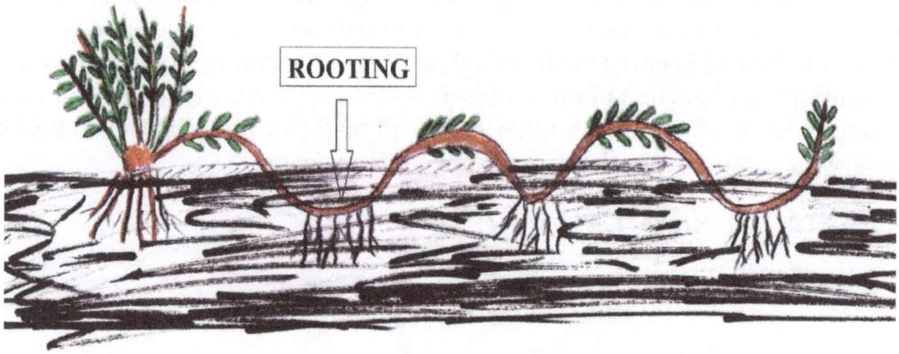

Fig 4.3: Compound or Serpentine Layering

Procedure

- ☆ Mature and flexible shoot of about one year shoot near to the ground of plant is selected.
- ☆ Remove the leaves retaining the leaves on the tip portion of the stem.
- ☆ Give the tongue cut at every spot (which to be burried) of the shoot.
- ☆ Apply the rooting hormones at the cut portion and cover with soil for early rooting.
- ☆ Stem is buried in the soil keeping the intermittent bud exposed.
- ☆ The shoot-initiated portion of shoot is ready for transplanting.

iii) Trench Layering

In this layering covered shoot in the soil result in etiolation so it is called as etiolation layering. After sprouting of new shoot and development of new roots from the buried portion it is ready for transplanting after separation from the mother plant *e.g.* apple, pear, cherry, walnut, rhododendron and jasmine.

Procedure

- ☆ Select flexible shoot of about one-year old which can touch ground easily.

- ☆ Detach leaves from lower portion and keep some leaves on the tip of the shoot.
- ☆ Make a shallow trench about 5-10 cm, place shoot in the middle of the trench and cover shoot with moist soil.
- ☆ Terminal portion of the shoot is kept exposed for synthesis of food and hormones for rooting and sprouting.
- ☆ After 2 to 3 weeks shoot arise from the buried shoot. The soil covered shoot result in etiolation and help in rooting.
- ☆ Sprouted and rooted shoots are ready for transplanting in the nursery.

iv) Mound Layering or Stooling

Mound layering is followed in the tress in which the branches near the ground level are not available for propagation or layering. In this method upper portion of plant at dormant stage is cut 2.5 cm above the ground level.

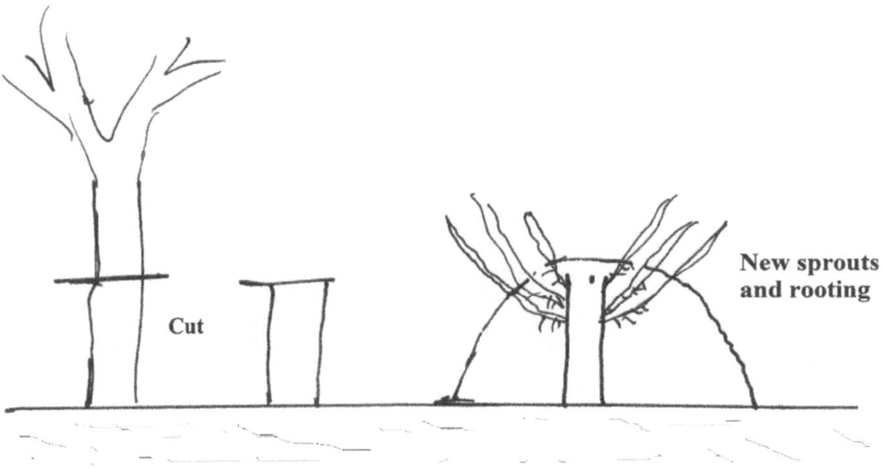

Fig 4.4: Mound Layering or Stooling

Procedure

- ☆ Cut upper portion of the plant about 2.5 cm above ground level during dormant stage.
- ☆ Few days after cutting shoot will emerge.
- ☆ When sprouts grow 7-15 cm cover the lower portion of the sprout with soil to initiate rooting.
- ☆ When sprout grow 20 to 25 cm then add some more soil to cover about 50 per cent lower portion of the shoots and apply water on soil covered shoots. It takes about 3 to 4 months to form a new plant. Separate these layered shoots from the mother plant for transplanting *e.g.* guava, apple, pear, gooseberry *etc.*

v) Air Layering

Formation of roots on the shoot attached to the mother plant above the ground level *e.g.* litchi, lemon, pomegranate, croton, *Callistemon*, philodendron, monstera *etc.* This rooted shoot is ready for transplanting in the nursery.

Procedure

- ☆ Select healthy and disease-free plant.
- ☆ Select the shoot of pencil thickness or about one year old and remove its leaves retaining on the tip portion.
- ☆ Remove phloem or girdle the shoot up to 2 to 3 cm wide between the nodes about 15 to 30 cm from the tip of the shoot.
- ☆ Scrap the girdled portion to remove the phloem and cambium portion for blockage of photosynthetic food for early rooting.
- ☆ Cover girdled portion with moist moss grass/sphagnum moss and wrap it with polythene sheet to retain the moisture on the girdled portion. Tie both ends of the polythene sheet air tight to prevent loss of moisture.
- ☆ Air layered shoot is removed from the mother plant and transplanted in the nursery. It takes about 4 to 6 weeks. It requires less time during rainy season.

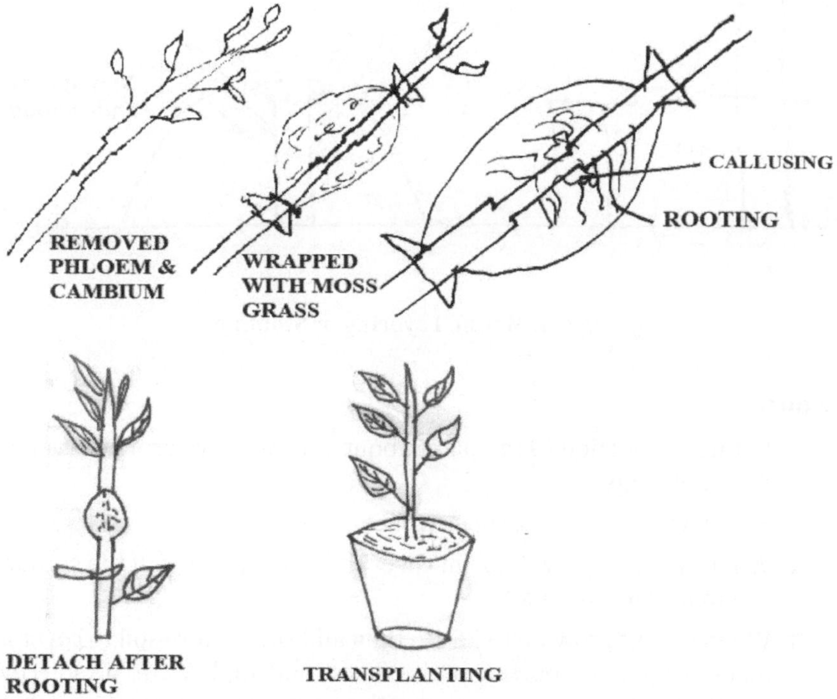

Fig 4.5: Steps of Air Layering

Fig 4.5 (a): Air Layering in Lemon

Plant Propagation by Grafting

Grafting

Two parts of plant *i.e.* rootstock and scion are joined together in a different manner to form a one functional plant is called grafting.

Advantages of Grafting

- ✰ Grafted plants bear flowers and fruits similar to mother plant and earlier as compared to seedling plant.
- ✰ This method is useful to multiply and preserve plants.
- ✰ Top working by using different grafting methods to improve the existing/ local variety with superior one.
- ✰ It can be used to repair the damaged trunk using bridge grafting.
- ✰ Influence of rootstock on scion like resistance against biotic (nematode, termite and soil borne diseases) and abiotic factors (drought and salinity/ alkalinity), vigour of scion and quality of produce.

Disadvantages of Grafting

- Success rate depends on the skill of person (skilled person is required).
- It is time consuming and expensive method of propagation.
- This method is not helpful to increase the genetic diversity, new variety or variability cannot be achieved by grafting.
- Plants propagated through grafting are less vigorous, short lived and mostly susceptible to insect pest and diseases.
- Some diseases may spread during grafting operations by propagation tools in newly propagated plants.

Rootstock

It is underground or root portion of the plant which gives support and supply water and nutrient to the plant on which scion is grafted. Most of the rootstocks are propagated by seeds in seed beds or polybags.

Characteristics of Rootstock

- Rootstocks should be adaptable to local soils and climate.
- Resistant to adverse soil related problems like salinity, alkalinity, drought; and, insect pest and diseases *etc.*
- It should be multiplied easily and compatible to scion.
- It should promote early healing of joint of stock scion.

Raising of Rootstock

Method of propagation of rootstocks varies with the crop, however most of these are multiplied by seeds *e.g.* ber, aonla, mango, citrus, guava, bael or sometimes cuttings pomegranate, mulberry, rose *etc.* Seeds or cuttings are grown/planted in raised nursery beds or polybags.

After germination of seeds or sprouting of cuttings, these are shifted in polybags for proper growth. Once these rootstocks attained pencil thickness or sufficient height, these are being used to insert scion portion to prepare a budded plant. Sometimes, the rootstocks of different species are used, *e.g.* grape rootstock of *Vitis berlandieri* is used for *Vitis vinifera*; *Zizipus rotundifolia* is used for *Ziziphus mauritiana.*

Scion

The vegetative part of elite type or improved variety taken from mother plant to be used as upper portion of graft combination for multiplication is called scion.

Characteristics of Scion

- Scion should be matured properly, most of the scion are taken from the growth of previous season.

- Collection of scions from flowering and recently harvested shoots should be avoided.
- Scion should be taken from well performing tree.
- Scion shoot should be well developed and healthy.

Selection of Scion

- Its produce or fruit should have good market value.
- It should be taken from grown up tree compatible to the rootstocks.
- Defoliate the branch about one week before the budding or grafting for swelling of buds.
- It should be taken form high yielding, true to type mother plant. It should be free from insect pest and diseases or viral diseases.

Methods of Grafting

i). Approach grafting/Inarching

ii). Tongue grafting

iii). Veneer grafting

iv). Side grafting

v). Wedge or cleft grafting

vi). Stone or epicotyl grafting

vii). Whip or splice grafting

viii). Bark grafting.

ix). Bridge grafting

i) Approach Grafting/Inarching

In this method rootstock or seedling plant is grown in pot or polybags, when this seedling attained pencil thickness, placed near the mother plant. Scion and stock, two independent plants approach together to join each other. These plants are not detached until the union takes place. This method of grafting is called approach grafting or inarching.

After a successful union of scion and rootstock, the rootstock portion is removed/detached above the graft union and scion is detached below the graft union to form a single grafted plant. This method of grafting is useful for the plants in which simple graft/bud success is very difficult. In this success rate is high because both scion and rootstock plants have their own food supply or photosynthesis system. However, in this method one can multiply limited number of plants from a single mother plant *e.g.* guava, mango, sapota.

Selection of Grafting Material

On the basis of cut at the joint portion the approach grafting is classified as:

- ✰ Sliced approach grafting
- ✰ Tongue approach grafting
- ✰ The success of grafting depends on the methods used and compatibility of rootstock and scion, age of the shoot, thickness of scion shoot/rootstock and season or environmental conditions. Scion should be elite type or improved one.

Fig 4.6: Approach Grafting/Inarching

Sliced Approach Grafting Procedure

- ✰ Place the rootstock near the mother plant.

- ☆ Find the most suitable contact point of stock and scion which have minimum chances of damage/break/stretch on each other.
- ☆ Remove slice of wood and bark 3 to 5 cm long with 1/3rd thickness of stock as well as scion.
- ☆ Place or bring the sliced portion of rootstock and scion together, these should be uniform without air space in between. There after tie them with budding tape or polythene. When rootstock and scion attached successfully cut the scion portion below the union point and rootstock above the union. Plant is ready for transplanting.

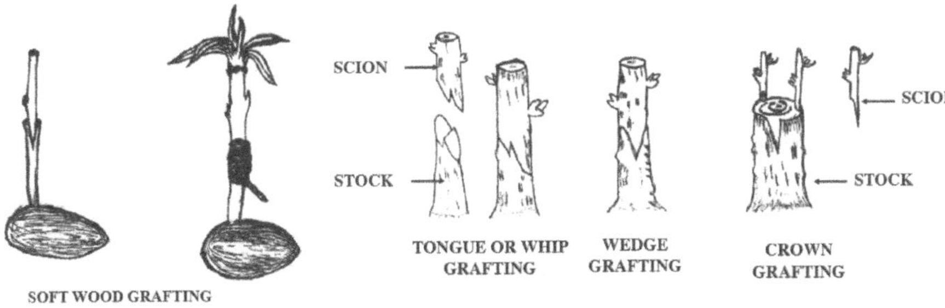

Fig 4.7: Different Methods of Grafting

ii) Tongue Grafting

In this method of propagation tongue like cut is given on the rootstock and scion plant to join them together *e.g.* apple, pear, walnut.

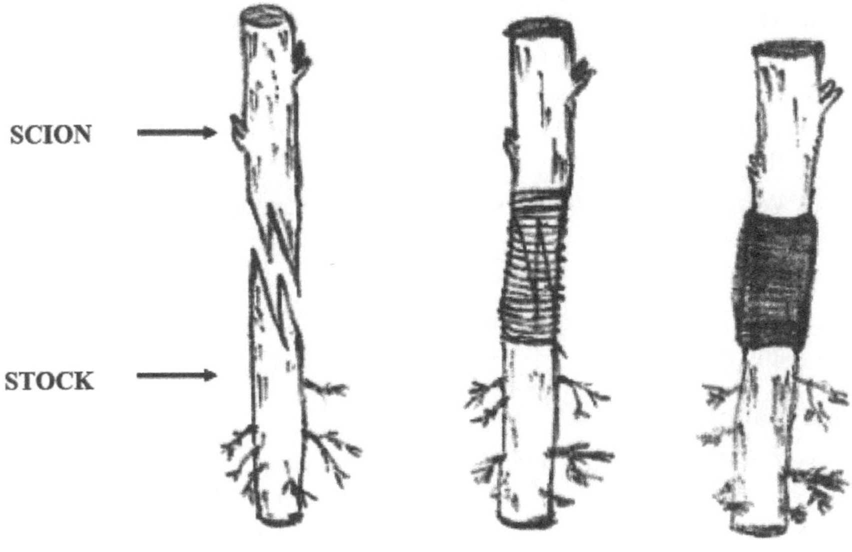

Fig 4.8: Tongue Grafting

Procedure

- Select the scion and rootstock of almost same age and thickness. Bring them close together.
- Select the point on rootstock for more successful attachment.
- Make a sliced cut of 3 to 5 cm long bark on rootstock and scion.
- Make a second slanting cut in the centre of the first, downward on the rootstock and upward on the scion cut in the form of tongue.
- Interlock the tongue by inserting the scion in stock.
- Operated portions should not be kept open or must be in contact with each other.
- Cover the operated portion with budding tape or polythene.

iii) Veneer Grafting

It is easiest grafting method. This method is ideal for top working of rejuvenated old and unproductive orchard and *in situ* grafting. In north India best time for veneer grafting is February-March and July-August. This technique is most commonly used in mango, cashew and peach.

Procedure

- A shallow downward cut of 3 to 5 cm long is made on lower portion of rootstock.
- At the base this cut make horizontal cut that intersects the first cut, remove the piece of wood and bark between both the cuts by making a small notch in the rootstock.
- The similar cut of same length is made at the lower end of the scion and small cut on the opposite side at the base.
- Insert or fix the scion into the rootstock. Ensure that at least half the cambium layer of the rootstock should be matched with cut surface of the scion.
- Wrap this union point of scion and stock with polythene strip or bud tape.
- After successful union/sprout cut back the rootstock portion above the union point.
- This method of propagation is used for grafting shrubs, deciduous trees and conifers.

iv) Side Grafting

In side grafting operated scion (shallow downward cut on lower side of scion) is inserted into the side of well-established rootstock of girth more than the scion, *e.g.* hibiscus.

Selection of Material

- ☆ Select a rootstock of about 2.5 cm diameter.
- ☆ The scion should be about 7.5 cm long with 3–5 buds.
- ☆ Rootstock should be comparatively thicker than scion.

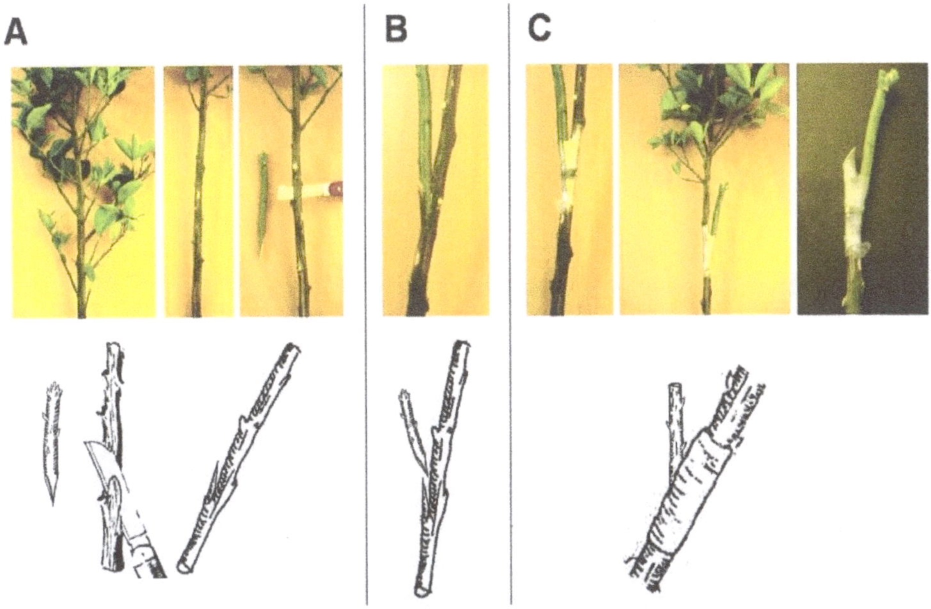

Fig 4.9: Side Grafting

Procedure

- ☆ On the lower end of the rootstock a downward slanting cut of about 2.5–5 cm deep is made.
- ☆ A wedge of same size cut (2.5–5 cm) on the lower end of scion opposite towards the base is made.
- ☆ Insert the scion into cut portion of the rootstock and pour wax to make operated portion water proof.
- ☆ Wrap or cover the operated portion or union portion of the stock and scion grafting tape/polythene sheet to keep it intact until they join.
- ☆ After successful completion of grafting separate the plant above the joint.

v) Cleft/Wedge Grafting

Wedge grafting is simplest and easy grafting method in fruit trees like mango, bael, jackfruit, aonla *etc.*

Fig 4.10: Fleft or Wedge Grafting

Selection of Material

- ☆ Select buds from current season growth.
- ☆ Remove leaves from the selected scion; it helps in accumulation of food and swelling of scion buds at about 2 weeks before the separation from mother plant.
- ☆ Scion thickness (diameter) may be same or less to the rootstock.

Procedure

- ☆ Raise suitable rootstocks in poly bags.
- ☆ Head back rootstock 4 to 5 months after planting.
- ☆ Make 3–5 cm downward cut in the centre of the stock.

- ☆ Two slanting cuts of 3-5 cm length on the sides towards the base of the scion are made.
- ☆ Split of the rootstock from the centre and insert the wedge-shaped scion in the cut portion.
- ☆ At least one side of the cambium of scion should match with the cambium of the rootstock while inserting the scion for better union.
- ☆ Cover the operated portion of scion and stock with budding tape or polythene tape.
- ☆ After successful completion of union or sprout of scion remove the polythene tape to allow the normal growth of the shoot.
- ☆ Support the newly grafted plant to avoid the breakage from union point.

vi) Stone or Epicotyl Grafting

This method is most commonly used in multiplication of mango plants. In this method, mango stone (seed) is sown in moist bed or polythene bags and covered them with layer of leaf mould of about 5 to 7 cm for better germination. Fifteen days after sowing seeding are grafted with epicotyl grafting method *e.g.* mango, tea.

Selection of Material

- ☆ Mango seedlings of about 15 days old having copper coloured leaves are used as rootstocks.
- ☆ Current season shoot with 4-5 active buds of 3 to 4-months old of pencil thickness is taken as scion.
- ☆ Defoliate the scion shoot about 15 days prior to collection of scions to make the buds swollen for early sprouting.

Procedure

- ☆ Make about 5 cm slanting cut at the base on both side of scion to make wedge shape.
- ☆ Head back the rootstock by giving a straight horizontal cut.
- ☆ Give 5 cm vertical cut downward from centre of the stock.
- ☆ Split the cut portion of stock and insert the wedge-shaped scion to allow full contact of rootstock and scion for easy union.
- ☆ Tie the union portion of graft firmly with budding tape or polythene strip. The sprout or emergence of new shoot is a sign of successful graft.
- ☆ Successfully sprouted plants are ready for transplanting within one year, *e.g.* mango.

vii) Whip or Splice Grafting

In this method thickness of the rootstock and scion are same. Select scion of 10 to 15 cm length with 4 to 5 active buds. Rootstock should also be in growing stage or sap flowing stage. This is oldest most commonly used method of grafting. This method is preferred in the early spring season *e.g.* apple, walnut, pear *etc.*

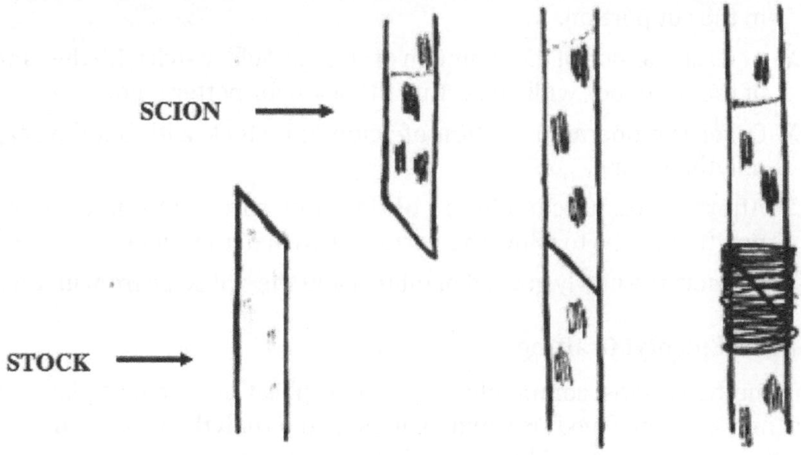

Fig 4.11: Wedge or Splice Grafting

Procedure

- First do head back the rootstock terminally at active growth stage.
- Give 2.5–5 cm long slanting cut from top to downwards.
- Slanting cut on scion of same thickness and same size is given from the base to upward.
- Cut on rootstock and scion should be smooth.
- Place the cut portion of scion on rootstock in the form of single stem.
- Wrap the union point with budding tape or polythene sheet.
- Remove the tape after successful union of rootstock and scion to avoid the restriction of growth at bud union point.

viii) Bark Grafting

It is method of grafting in which the scion is inserted by slitting the bark. It is most commonly used in top working.

Selection of Material

- The bark of the plant to be recovered must be in active phase or sap flowing conditions.
- The scion/shoot to be inserted must be in dormant condition.
- The scion should have 3 to 5 dormant buds with 10–13 cm length.

Procedure

- 3 to 5 cm long vertical cut in the bark of the stem/rootstock is made.
- 3 to 5 cm long sliced wedge shaped cut on both side of the shoot should be made.
- Loosen the cut portion of bark of rootstock to insert the shoot/scion.
- Insert the wedge-shaped scion in the loosened portion of the rootstock beneath the bark and apply wax on exposed portion to make it water proof.
- In the same way several scions should be inserted in a single stock and number of scion or shoots depend on the thickness of the stem/trunk.

Grafting for Special Purposes

This type of grafting is followed to improve the quality of the existing plant. In this method the existing plant is headed back and the new sprout from the headed back portion are used as a stock for grafting of improved variety. This can be done with side grafting and top working. It is followed to rejuvenate the old orchard or injured tree.

ix) Bridge Grafting

Grafting is done on the girdled bark to recover the stem damaged by implement, frost, disease and rodent. In this grafting the shoots/scions are inserted into the stem above the damaged portion and similarly below the damaged portion to overcome hindered supply of nutrients and water due to damaged portion. It is done to bridge the girdling gap.

Selection of Material

- The rootstock should have sufficient sap flow for successful grafting.
- Scion shoot should be about one year old and of pencil thickness.
- Number of scion sticks used depend upon area of the trunk or wound to be repaired.
- The scion should be from the compatible plant or if possible, it should be taken from the same plant.

Procedure

- Remove the dead bark by trimming the wound area.
- Cuts are made below and above the wound in the bark to insert the scion at 5 to 7.5 cm distance.
- Long slanting cuts on the scion are given to increase cambium attachment area and insert easily below and above the wound into the stock.

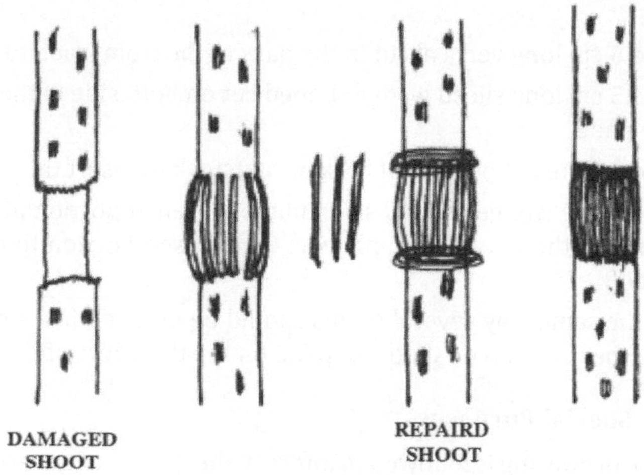

Fig 4.12: Bridge Grafting

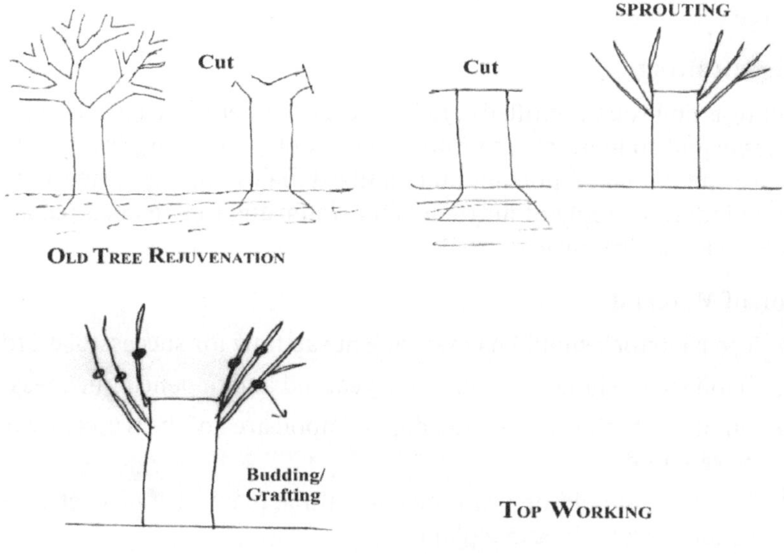

Fig 4.13: Steps of Top Working

- ☆ A small additional slanting cut opposite to the first cut at both the end of the scion is given to form a sharp wedge of scion.
- ☆ Make similar type of cuts on required number of scions.
- ☆ Take buds from scion(s).
- ☆ Insert the scion slots of bark.
- ☆ Put the scions in upright position only and wax each graft unions from top to bottom.

Top Working

It is the method of grafting in which inferior, older or senile plants are rejuvenated into elite/superior or new ones. This method is less costly as compared to uproot and new plantation. It is suitable for most of the fruit plants like aonla, mango, apple, citrus fruits, avocado and shrubs or vines. Top working is done by grafting/budding required scion on newly emerged shoots of rejuvenated plant. In top working according to the suitability of plant different methods of grafting or budding like cleft, whip, wedge or side grafting methods can be used. Usually it is done during spring season.

Fig 4.14 Rejuvenation of Ber Plant

Selection of Material

- In this method the rejuvenated plant will act as rootstock.
- Top working is done on 3–5 scaffold branches of rejuvenated plant.
- Secondary scaffold branches are grafted for frame working.
- Top working should be done on branches well distributed around the trunk.
- The branches used as stock should have thickness about 3–10 cm.
- Scion sticks should have swollen bud.

Precautions

- ☆ Check or observe the progress/success of the top worked branches at every 3–5 days.
- ☆ Re-waxed on the cracks developed in the wax coating to increase the success.
- ☆ Sun burn can be avoided by white wash during summer.
- ☆ Protect the newly grafted scion from high temperature or sun light by covering the graft with shade net.
- ☆ Newly emerged shoots from the graft union should be tied to avoid the breakage due to wind from the graft union.
- ☆ Growth or sprouts other than grafted portion from older branches and trunk must be removed from time-to-time.
- ☆ The top worked tree should be irrigated at regular interval and also manured as and when required.

Budding

Budding is the process in which single bud is inserted into a rootstock that results into union of scion/stock, which grows as a new plant. It is also grafting but single bud is used as scion.

Types of Budding

Different methods are used to insert bud into the rootstock. Some of the most commonly used methods are as follows:

i) T-Budding/Shield Budding

'T'-shaped cut is made on the rootstock for bud insertion is called T-budding. It is also called as 'shield budding' as the shield shaped bud is inserted into the stock. It is most common method used for propagating different fruit trees and ornamental plants. In this method of propagation, the compatible rootstocks are raised in beds or poly bags. One-year old healthy and vigorous rootstock of pencil thickness is selected for propagation. For a T-shaped cut, first make the horizontal cut on the bark of rootstock at about 15 -25 cm above the ground level after that downward vertical cut 2.5 to 3.5 cm long from the centre of the horizontal cut is made in such a way that it look likes T-shape with the help of a sharp knife. Collect the shield-shaped swollen bud from the desirable mother plant (true to type) with the help of sharp knife. Loosen the bark of rootstock from the T-shaped cut portion to insert the bud. The shield bud is inserted in the T-shaped cut in the stock. After the insertion of the bud into the cut portion of the rootstock with the help of a budding knife, the bud inserted portion should be firmly wrapped with budding tape or polythene strip by keeping the bud exposed. The bud sprouts will start within a month after that plant is ready for transplanting. This method of propagation is followed in ber, rose, pear, peach, cherry, apricot, sweet orange *etc.*

Fig 4.15: T-budding/Shield Budding

ii) Patch Budding

A rectangular shaped patch (2.4 cm ×1.5 cm, length and width) of bark between internodes of the stem is removed from the rootstock plant. The patch size of scion and rootstock should be same. The patch along with healthy bud is placed on the cut portion of the stock keeping the bud upward and wrapped with a polythene strip, keeping the bud exposed. This method of budding is used for propagation of plants having a thick bark *e.g.* aonla, jamun, citrus, rubber *etc.*

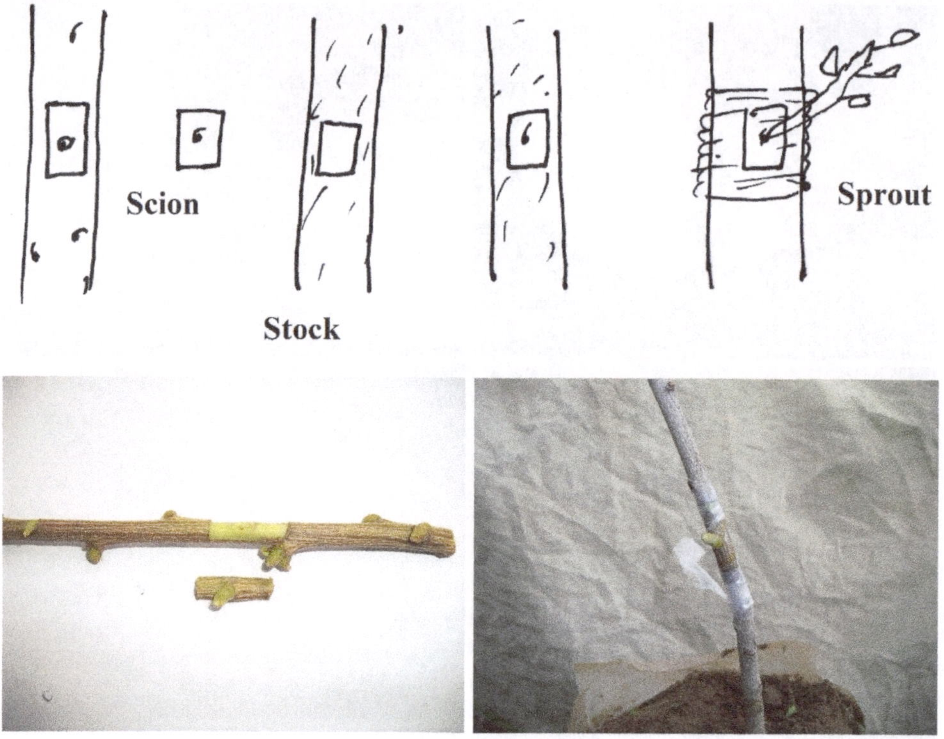

Fig 4.16: Patch Budding

ii) Ring Budding

In this method a circular bark of about 3 cm height in ring form is removed from the rootstock. The bark of the same width scion along with healthy bud is removed from the mother plant and placed on the rootstock. After placing the ring on the cut portion of the rootstock, tie it with a budding tape. This is most successful in ber and cherry.

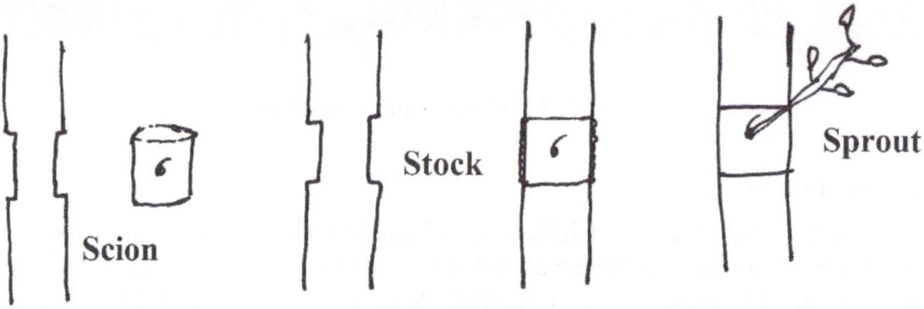

Fig 4.17: Ring Budding

iv) Flute Budding

Flute budding is a slightly modified ring budding. In flute budding instead of removing complete ring, 1/8th portion of ring kept attached to the rootstock. A similar portion of the scion is removed along with bud from bud stick of the mother plant and fit it on the cut portion of the rootstock. After fitting of budded bark on the root stock it is tied with the polythene strip, keeping the bud point exposed *e.g.* ber.

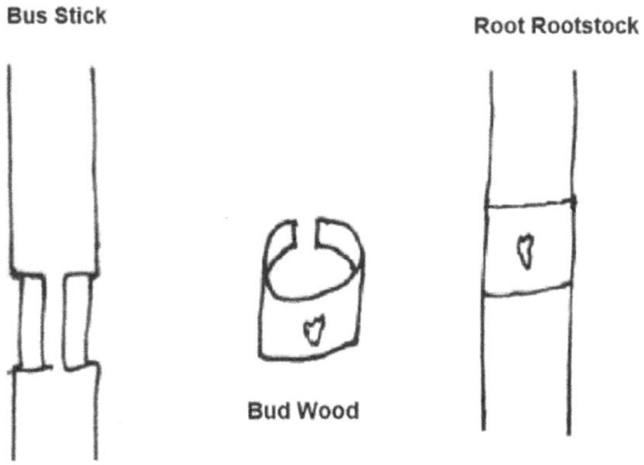

Fig 4.18: Flute Budding

v) Forkert/Flap Budding

In this method horizontal cut between the internodes of the rootstock is given at about15–25 cm above the ground level. Thereafter two vertical downward cuts from the cut ends of the horizontal cut is given, a flap of the bark is pulled downward and kept in a way that lower portion attached to the stock. The bark exposing portion is about 2.5 cm × 5 cm. A rectangular piece of bark along with active bud is taken from the scion wood bud stick with the help of sharp budding knife. The scion piece of bark is then fitted on the exposed bark portion of the rootstock. Then the flap of the bark is tied to its previous position covering the patch of scion.

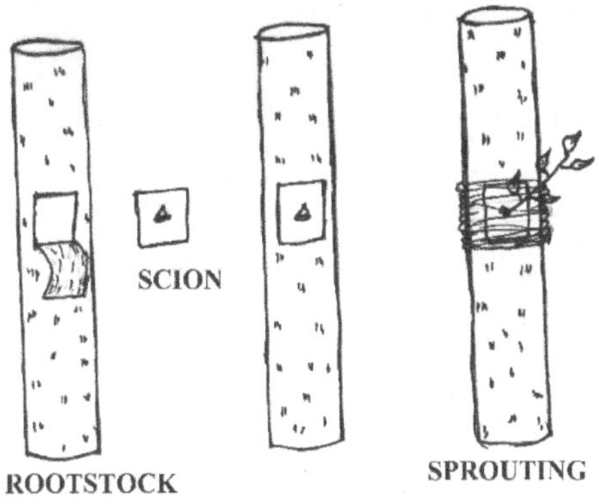

Fig 4.19: Forkert/Flap Budding

Observe three weeks, either sprout will pull out the flap or if bud inside the flap is swollen than remove the flap by giving the horizontal cut at lower portion of the flap. If bud start sprouting than wrap the polythene tape again keeping the bud/ growing point exposed. Sprouting of bud takes place within 3 – 5 weeks of budding operation *e.g.* mango, jackfruit and cashew nut *etc.*

vi) Chip Budding

Chip budding method of propagation is followed in the crops which have very thin bark and it is very difficult to remove the bark. So, the bark of rootstock is removed along with the wood portion and the bark with wood of same size and shape of scion is removed. Place this scion wood bark on the cut portion of the rootstock. This method of propagation is done in the month of February- March in crops like apple, pear and grapes.

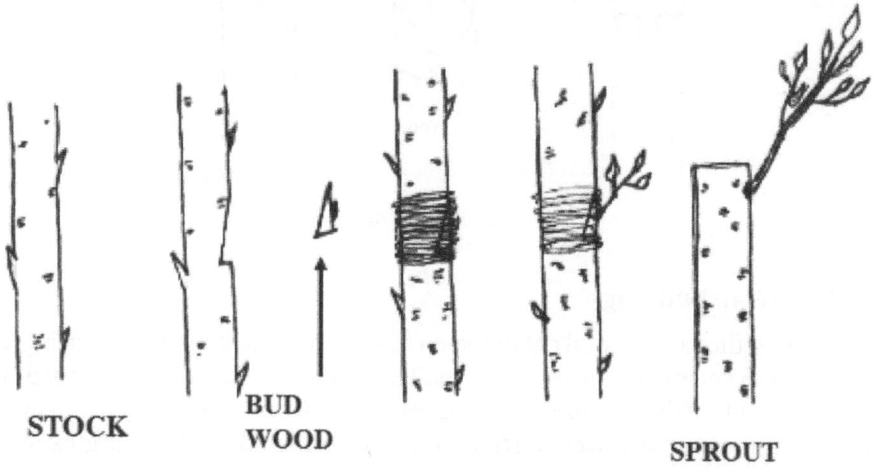

Fig 4.20: Chip Budding

Micropropagation/Tissue Culture

This method of propagation is followed when we need to propagate large number of plants in short period. In this method large number of disease-free plantlets can be prepared in a tissue culture laboratory. In this technique plant tissues are isolated from the parent plant and multiplied in the artificial medium under controlled environment under aseptic conditions. This method of multiplication is done in banana, papaya, gerbera, rose, carnation, orchid, anthurium *etc.* The multiplication is based on 'totipotency' phenomenon of a cell; means ability of cell to regenerate into plant. Callus is produced on explants under *in vitro* conditions due to endogenous or exogenous growth substances. The explants can be collected from different parts of the plant such as root, stem or leaves. After disinfestations the plant parts are induced in the growing medium to form 'callus'.

This technique eliminates systemic diseases from the plant materials and provides healthy and uniform plants to the growers round the year.

Advantages of Micropropagation

- It eliminates systemic diseases from plant materials using disease eradication techniques such as meristem culture or meristem tip culture.
- Reduce the space required for maintaining rootstocks and propagated plants, and provides disease free healthy plant of uniform size to the grower round the year.
- This method of multiplication is helpful for propagators to select desirable traits/variants and generate into new viable cultivars easily and quickly. More than 80 commercial cultivars have been selected from variants and introduced into the trade.
- It also speeds up the introduction of hybrid cultivars. Newly released cultivars can be multiplied into sufficient numbers of plants through micropropagation and it becomes available within 2 to 3 years, however, it takes 5 to 7 years through traditional method of propagation.

  Stage 1, establishment of clean cultures;

  Stage 2, multiplication;

  Stage 3, root initiation; and

  Stage 4, acclimatization

Propagation by Specialized Plant Organs

Specialized plant organs are modified plant parts such as leaf, stems or roots, developed above the ground surface or underground, which are used for multiplication of plants. Most common methods are explained below:

i) Bulb

It is specialized underground portion having a flat basal stem and surrounded by fleshy scales, *e.g.* tuberose, onion, amaryllis. Structurally bulbs are of two types

i) Tunicated bulbs: The outer layer is converted into dry membranous for protection of the bulbs, *e.g.* onion, tuberose, amaryllis, tulip *etc.*

ii) Non-tunicated bulbs: These bulbs are not covered by dry covering *e.g.* lilium.

ii) Corm

Corm is an underground modified solid or compressed stem oriented vertically in the side having nodes and internodes, *e.g.* gladiolus, crocus *etc.*

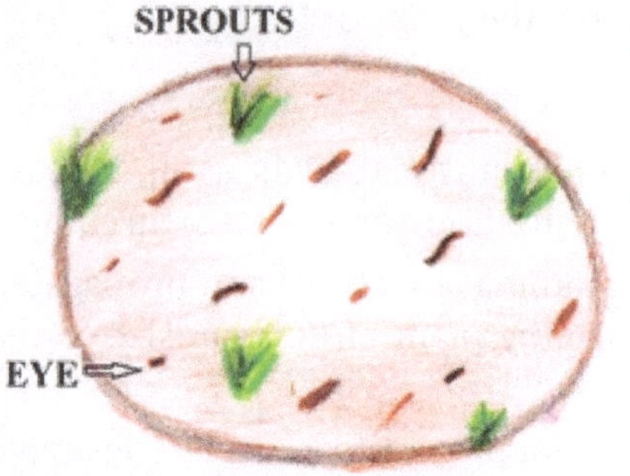

Fig 4.21: Tuber

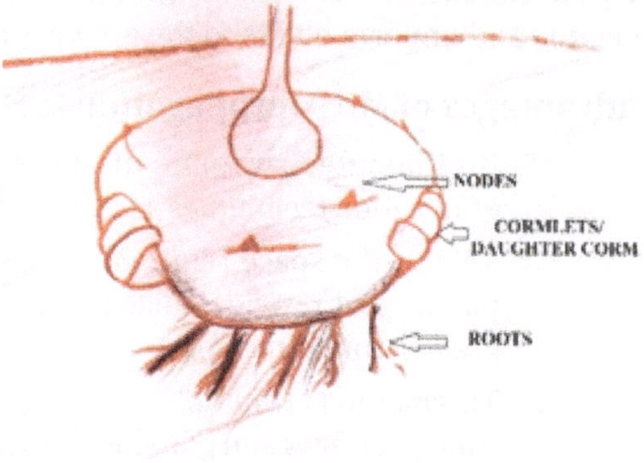

Fig 4.22: Corm

iii) Tuber

Underground storage organ of plant having special swollen modified stem or roots.

- ☆ Root tubers *e.g.* dahlia, caladium, dioscorea, Jerusalem artichoke *etc.*
- ☆ Stem tuber *e.g.* begonia, potato *etc.*

iv) Rhizome

Underground horizontally growing modified stem just below the ground surface is called rhizome, *e.g.* canna, ferns, ginger, iris *etc.*

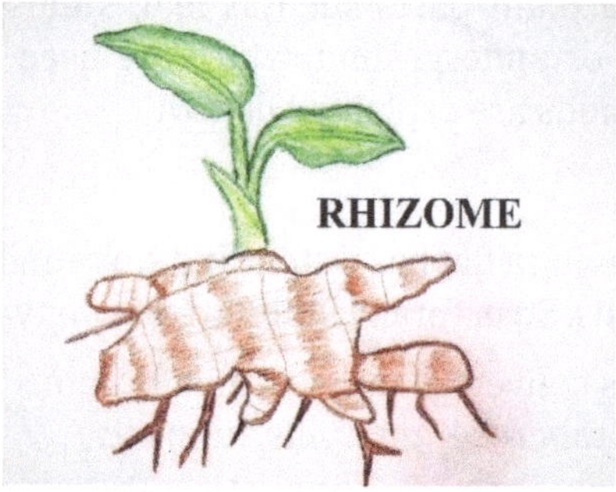

Fig 4.23: Rhizome

Fig 4.24: Runner

v) Runner

Horizontally growing creeping nature modified stalk emerged from the leaf axil and produce root on each node. It can be used to produce separate or new plant, *e.g.* doob grass, strawberry, chlorophytum, *etc.*

vi) Sucker

The shoot or root arise from the root (just below the ground level) or stem portion (just above the ground level) of a plant are called suckers, *e.g.* chrysanthemum (stem), *Clerodendrum splendens* (root suckers), anthurium *etc.*

Fig 4.25: Sucker

Fig 4.26: Tissue Culture/ Micropropogation

vii) Tuberous Root

Swollen tuberous growth act as a storage organ. Examples are banana, date palm, pineapple, satavari, dahlia, chlorophytum *etc.*

Chapter 5

Plant Propagation Structures

Plant Propagation Structures

The structures used to increase the success rate of propagation are called propagation structures. These structures are used for multiplication of plants by different propagation methods such as seed, layering, inarching, cutting, budding and grafting. Propagation structures are special structure and equipped with necessary equipments to provide optimum environmental conditions for normal physiological growth activity to the nursery plants.

Environmental Factors Affecting Propagation

During propagation the appropriate environmental conditions must be provided for successful or more chances of success during propagation in lesser time because the sun light, air, temperature and humidity are important factors which influence the success in propagation. The nutritional conditions of the growing media is also one of the important factors. The young plants also need a protection form insect pest and diseases during the propagation and growing media related problems like problematic soils.

Light: It is a source of radiant energy and important for photosynthesis. The management of light can be critical for rooting, cutting, germinating seeds, growing seedlings, or shoot multiplication of explants during propagation. Light can be manipulated by controlling irradiation, light duration (photoperiod) and light quality (wavelength).

Humidity: The mist or fine droplets of water are provided to decrease the temperature and increase humidity. It reduces the heat load and creates moist and cool microclimate in the surroundings.

Temperature: Temperature affects propagation in many ways.

1. Stratification: Allow the germination process.
2. Rooting: Ambient air temperature or cooling effect of mist.
3. Grafting: Some time high temperature required in graft union area.

Gases and gas exchange: The balance of O_2 and CO_2 is required during different process of propagation and it can be maintained in the propagation structures.

- During propagation in enclosed green house, ambient CO_2 levels can drop to suboptimal level which limits photosynthesis and propagules development.
- Buildup of ethylene gas can deteriorate the propagules during storage, propagation and shipping.

Mineral nutrition: Application of nutrients in the form of soluble fertilizers during irrigation of a seedling or root cutting.

Pests and diseases: Soil less media to avoid soil borne diseases.

Propagation Structures

i) Greenhouses

These are structures covered with plastic or glass. These structures allow the short-wave radiations coming from the sun but do not allow the long wave radiations reflected from the ground surface. This increase in temperature is utilized for propagation during winter season.

Fig 5.1: Poly House Structure

- ☆ During summer season the temperature can be controlled or lower down by using cooling pad and fan.
- ☆ Higher CO_2 concentration inside the green house is helpful to increase the plant growth by improving the photosynthetic efficiency of the plant.
- ☆ In addition to this the humidity can be increased or decreased inside the greenhouse as per requirement of the plants according to their growth stage.
- ☆ Light requirement of the plants can also be adjusted in the green house by providing artificial light and shade.
- ☆ The requirement of irrigation and nutrients can also be managed easily.

Greenhouse on the Basis of Covering Material

Plastic covered: Covered with the plastic sheets to lower down the cost of erecting of the structure and these are light in weight. Mainly two types of the plastics sheets are used for erection of the poly house construction are polythene and fiberglass. The life of the plastic covered structure or poly house is less as compared to the fibre glass covered structure.

Glass covered: Glass covered greenhouse structures are expensive but have long life.

Classification of greenhouse on the basis erection cost:

- ☆ **Low cost greenhouse:** No specific control system, structure is covered with ultra violet (UV) stabilized film as the cladding material.
- ☆ **Medium cost green house:** May have double layer of UV stabilized film with cooling pad/fan or heating requirement.
- ☆ **High cost green house:** UV stabilized low density polyethylene (LDPE) film (200 micron)/polycarbonate sheet, humidification, drip system, auto control system.

Advantages

- ☆ Plants can be raised during off season, so it decreases the duration for propagation.
- ☆ Quality planting material can be raised and it will be more economical.
- ☆ Plant can be easily raised under harsh environmental conditions.
- ☆ Plants can be raised or propagated under adverse soil conditions.
- ☆ More number of plants can be raised or propagated per unit time and per unit area.
- ☆ Pest and disease-free plant can be prepared

ii) Hot Frames (Hot beds)

It is used to increase the temperature of soil in the bed by using different manures, electric steam or hot water pipes. Such types of beds are covered with the glass or plastic frame to raise the temperature by greenhouse effect. Such types of beds are used for raising seedling and early rooting in cuttings. It is cheap and easy method for raising nursery plants without erection of large and costly structure. Hot beds are used for forcing plants or for raising early seedlings.

iii) Cold Frames

A bottomless box with a removable top is used for cold frame. In this type of frame artificial heat is not required but the temperature inside the box increased during the day by sun heat and keep the plant warm during night hours. This frame can be covered with the straw/grasses or crop residue. It is used to protect small plants from wind and low temperature. No artificial heat or manure is used inside a true cold frame. These frames are used for hardening of the rooted cuttings and young seedlings before transplanting in the open field. These are most commonly used to raise seeding during the late summer. Irrigation or mist in the cold frame is needed to maintain the humidity. High temperature can be controlled by ventilation and shading.

iv) Lath Houses

Lath (lath means thin strip of wood) house is framed with the wooden strips to protect the plants from high temperature and high irradiations during summer. It also reduces the moisture stress and water requirement of the plant. These are used in propagation and for hardening of the plants before transplanting in the open field. During mild temperature these can be used for propagation of the plants.

v) Net House

It is most widely used propagation structure. This structure is being used in tropical and subtropical conditions. It is used when the artificial cooling is expensive and heating is not required. The roof of net house are made up of plastic or glass while the sides are covered with net.

Fig 5.2: Net House Structure

Purposes

- ☆ Lower down the temperature
- ☆ Prevent the entry of insects
- ☆ Control the sun light

Advantages

- ☆ Off season cultivation can be done easily.
- ☆ Avoid the entry of insect pest and vector.
- ☆ It is helpful to increase the quality of the product.

Type of Net House

a) Shade Net House

Such types of net house are used to protect the plants or seedling form the high temperature or high irradiation during the summer months. These structures can be supported with wood or metal pipes. The cladding material of shade net is available in different colour and allows shade in different percentage. Generally, these are used to harden the seedlings or cuttings and used for proper growth during high light intensity especially in tropical and subtropical areas. These are also helpful to grow shade loving plants.

b) Insect Proof Net House

These are specially used to prevent the attack of insects inside the net house; these can be framed as permanent as well as temporary structures. Insect proof net of 40-50 mesh is effective to control the insect-pest and diseases. Structure can be fabricated in different shape such as dome shaped and flat roof shade. In these structures natural or organic crops can be grown by avoiding the spray of health hazardous chemicals.

vi) Mist Propagation Unit

Spray of fine particle of water to increase humidity and decrease temperature in growing area. Intermittent misting reduces the transpiration losses which help in rooting of softwood leafy cutting. This technique is now widely used by nurserymen. The aim of misting is to maintain humidity by a continuous film of water on the leaves, thus reducing transpiration and keeping the cutting turgid until rooting take place. Misting frequency depends on the ambient temperature and types of propagation and planting material.

Advantages

- ☆ Planting material can be prepared throughout the year.
- ☆ It reduces the loss of moisture from the plant body (transpiration).

Fig 5.3: Mist Propagation Unit/Chamber

vii) Growing Room

A building is constructed and artificial light is provided for the growth of the plants. In this room seedling are grown in tray or pots which are kept on bench, these benches are kept in a multistory system to accommodate more trays or prepare more seedlings per unit area. Fluorescent tube of 125 watt is illuminated on the trays to provide the light. Number of fluorescent tubes depends on the requirement of light for proper seedling growth. These tubes are also helpful to maintain the temperature inside the growing room.

viii) Bottom Heat

Mist chamber sometimes decline the temperature in the bottom of the cuttings so to increase the temperature in the rooting area of the plants soil warming cables are installed. These are placed about 5 cm below the rooting medium. It increases the rooting at fast rate and thermostats are installed to maintain or control the temperature.

Chapter 6

Orchards Management

Planning an Orchard

A careful planning is required before establishing an orchard because it is a long-term investment. Mistake committed initially on selection of location, site, planting distance, varieties and the planting material used may considerably reduce the returns on the investment, besides wastage of time, money and energy. Therefore, it is advisable to consult the Directorates, Department of Horticulture, Agricultural/Horticultural Universities/ICAR centres, Regional Research Stations, Krishi Vigyan Kendras before planning an orchard.

- Orchard is a long-term business so the planning should also be done with the guidance of the expert.
- Planning should include the climatic conditions, soil types, sub soil types, irrigation source, planting material/variety and pollinator (if required) and their availability, planting distance, marketing area and market demand.

Orchard is a piece of land used to grow fruit trees by utilizing various resources like land, water, trees and inputs. These resources should be utilized efficiently to increase the quality production on sustainable basis without adverse effect on the surrounding environment.

Principles for Planning the Orchard

- It should be located at a place where suitable climatic and soil conditions are favourable.

Wind Break: (Covered in Chapter 3)

Building: It should be planned before the planting, though their construction may be done later on.

II. Soil (Covered in Chapter 3)

Common soil management practices: (Covered in Chapter 3)

III. Topography

Flat and with slight slope, allowing good runoff from heavy rains. In general, the southern slope promotes early blooming and early maturity of fruits.

IV. Availability of Irrigation Water

Application of water at right time helps the plant to grow and bear properly. High level of chloride and sodium in irrigation water should be avoided. Boron salt more than one ppm should not be used for irrigation *e.g.* litchi at fruit development stage should be irrigated frequently.

V. Availability of Finance and Labour

Fruit growing is more labour and capital-intensive enterprise compared to field crops *e.g.* irrigation systems, application of inputs and harvesting.

VI. Availability/Nearness to Market

Most of the tropical fruits like mango, litchi, guava or jamun are difficult to store. Their edible qualities steadily deteriorate under prolonged cold storage. Nearness to market maintains quality and increase profit.

VII. Transport Facilities

Good transport facility and cold chain from site of produce to site of consumption may decrease the post-harvest losses.

VIII. Availability of Fruit Processing Unit

Due to high moisture content the fruits are having very short shelf life. Processing of the fruits increase the value of the fruits as well as decrease the post-harvest losses. Processed fruits can be utilized in a better quality during the off seasons.

IX. Pests and Disease Problems

In north India vegetative as well as reproductive malformation in mango. *Fusarium* wilt in guava, banana is also susceptible to wilt.

2. Selection of Fruit Crops Species and their Cultivars

Selection of right crop/variety to suit the planting conditions and situation is one of the major prerequisites for successful orchard establishment *e.g.* tropical fruits cannot withstand temperature below freezing point and fruits like banana suffering from chilling injury at temperature below 10°C. Subtropical fruits can tolerate subfreezing temperature for a short period. Generally, they require cool weather for growth and fruit development and damaged by temperature below 5°C. Temperate fruits plants are winter hardy. Some plants can tolerate temperature as low as -34°C if properly hardened.

It is better to grow more than one crop in your orchard because growing of single crop is risky it may be destroyed due to climatic hazard or certain other reason (pest and disease).

Consider the following points for selection of crops:

- Crop should have good market demand
- Better quality and productivity.
- Regular bearer.
- Select compatible cultivars with more yield.
- Provide satisfactory return.
- Pollinizer should be planted.
- Less infestation of pest, disease either soil borne or from neighbouring orchard and nematode.

Selection of Cultivar

Sufficient attention must be paid to the choice of cultivar, rootstock, morphology, yield and quality of fruits.

Points to be kept in mind before purchasing plants:

- True to type plants.
- Well adapted and acclaimed for high yield and good quality.
- Collect information about the performance of new and untried cultivars.
- Select vigorous, healthy and disease-free plant, which can bear transplanting shock.
- Age of plants should be between 1-1.5 years.
- Confirm about the rootstock in budded/grafted plants.
- Avoid plant having too much branching at bud union and very tall grafts.
- Plants having twisted, abnormal, deformed roots should be rejected.

Growing Single Variety of a Crop

Selection of single variety of a crop *e.g.* Alphanso variety of mango, Bhagwa variety of pomegranate, NA 7 variety of aonla and Gola variety of ber.

Advantages

- To maintain the purity of a variety.

- ✰ Planning and management/operation become easy.
- ✰ Harvesting at a same time reduce the cost.

Disadvantages

- ✰ Incompatibility leads to poor fruit setting and yield.
- ✰ Single variety may be susceptible to insect-pest and diseases.
- ✰ Sometimes a variety may bear irregularly *e.g.* Alphonso.
- ✰ Income once in a year difficult to run the regular expenses.
- ✰ Employment for labour is not regular.
- ✰ Variety may not satisfy the end user.

3. Layout, Planting Systems and Density

Before taking up the actual planting it is necessary to decide the ideal number of plants per hectare. Trees must be set at optimum planting distance to obtain the most efficient and profitable use of land. In other words, the primary objective is to develop maximum bearing surface in minimum time and space. Before adopting adequate spacing for a fruit, one should have proper knowledge about growth characteristics and morphology of the scion, kind of rootstock, climate of area, depth of soil and fertility management practices to be followed in the orchard.

Preparation of Land

- ✰ Properly cleaned land becomes easy for movement of men and machinery.
- ✰ All unwanted plants should be removed.
- ✰ Uncultivated land should be leveled and ploughed deeply.

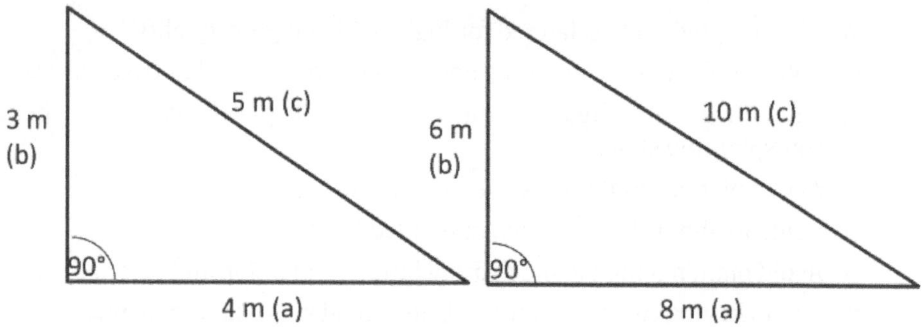

Pythagoras theorem : Relation between 3 sides of right triangle

$a^2 + b^2 = c^2$; $\sqrt{a^2 + b^2} = c$

$(3)^2 + (4)^2 = (5)^2$ or $(6)^2 + (8)^2 = (10)^2$

$9 + 16 = 25$ or $36 + 64 = 100$

$\sqrt{3^2 + 4^2} = 5$ or $\sqrt{6^2 + 8^2} = 10$

Fig 6.1: Procedure to Make Right Angle

Layout Plan

- Marking the position of plant in the field in called layout.
- First mark the position of the plant or layout in the field.
- Layout should be consulted with the expert.
- Layout should include the planning of the building, path, store and water channels.
- Sketch the plan on the paper before actual planting in the field.
- Try to accommodate more number of plants in available space.

Method of Layout

- First to form the base line for planting of trees in the orchard, plant a row of plant along the base line with the help of stick or rope.
- Form the other base line at right angle to the first base line.
- Make the right angle in the orchard with the help of measuring tape using Pythagoras theorem *i.e.* $c^2 = b^2 + a^2$. e.g. $c^2 = 4^2 + 3^2$, the value of c will be 5. In such a way we can make a right angle in the field.
- One end of this tape is fixed at three meters distance from the corner along the first base line and four meters towards 2nd base line and then diagonal distance between these two points should be five meters.
- The wooden sticks are put in the orchard area at the desired distance at crossing line point.
- Three men should stand in a line and the neck of third person should not be visible to first person while doing layout in the field for marking the position of plant.

Planting Systems

i) Square System

In this system of planting the distance between plant to plant and row to row is kept same. It is most commonly and easiest method of planting. The plants are planted at right angle to each other and unit of four plants form a square so it is called square system of planting. The intercultural operation can be done in both the directions.

Advantages

1. It is most popular and easiest method of planting.
2. Distance from plant to plant to plant is kept same.
3. Planting is done at right angle to each other.
4. Helpful for inter-culture operations in both the directions.
5. Inter-cultivation of vegetables can be done for additional remuneration.

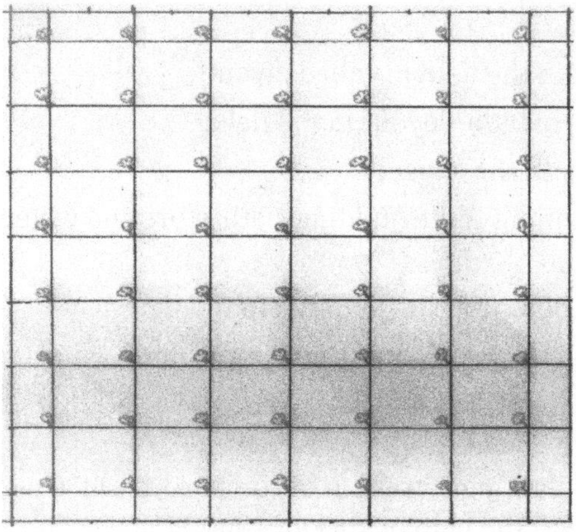

Fig 6.2: Square System

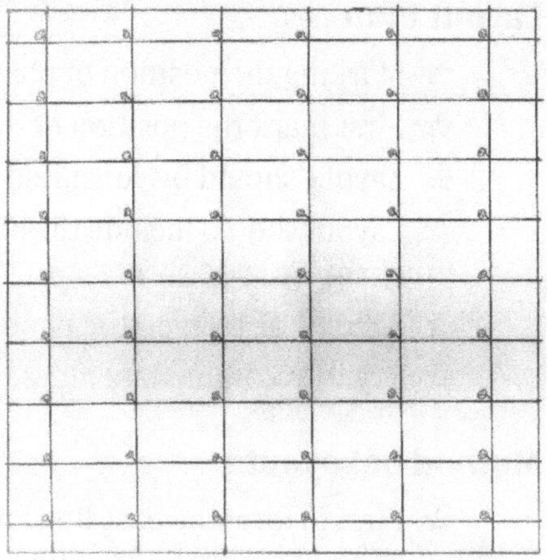

Fig 6.3: Rectangular System

ii) Rectangular System

In this system of planting unit of four plant make rectangle so it is called as rectangular system of planting. Intercultural operations can be done from both the directions like a square system. The only difference is that the space between the row is more and the space between the plants is less so more plants can be accommodated in the row.

Advantages

1. Lay out is made in rectangular shape.
2. Space between the rows is more for growing intercrop.
3. Inter-cultural operations can be done easily in both the directions.
4. Plants get proper space for spread, aeration and sunlight.

iii) Quincunx or Diagonal System

In this system planting is done in the square system and one additional tree of less height is planted in the centre of each square termed as filler plant. Filler plants planted in the centre of the square are phalsa, papaya, plum, peach, lemon, guava *etc.* It provides additional income to the grower. Filler plant can be uprooted when the main crop covers the whole space or it gave shading effect to the filler crop. It accommodates 78 per cent more plants.

Advantages

This method accommodates 78 per cent more plants.

Disadvantages

1. Inter-culture operations are not easy
2. Difficult to lay-out

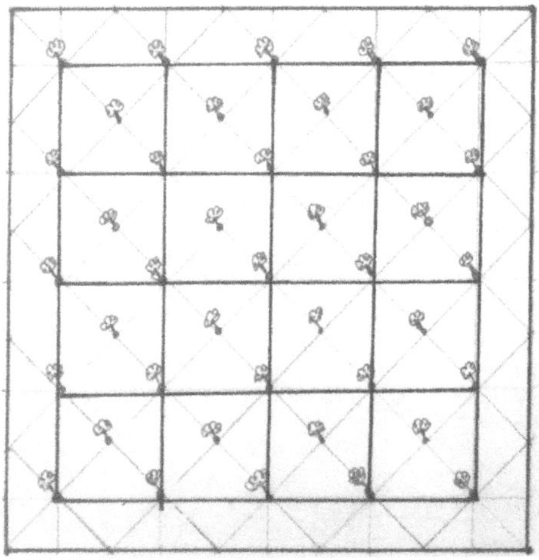

Fig 6.4: Quincunx or Diagonal System

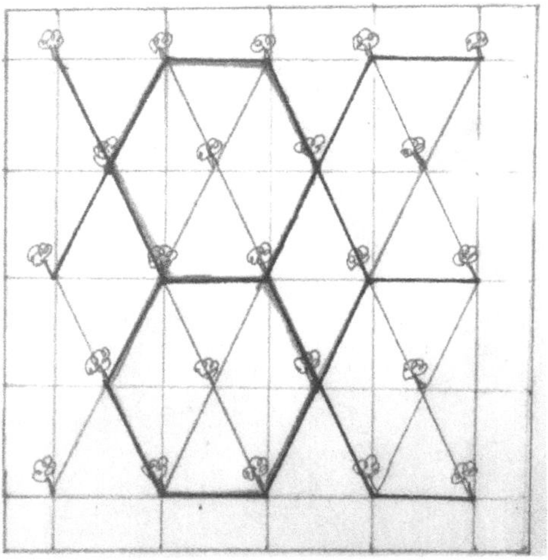

Fig 6.5: Hexagonal System

iv) Hexagonal System

The diagonals of rectangle are intersected to form equilateral triangles. Thus, trees form hexagonal with the seventh tree planted in the centre of hexagonal. Hence this system is also called as septule. In hexagonal system of planting trees are planted at the corners of equilateral triangles. This system of planting can accommodate 15 per cent more plant but difficult to plan. This system can be followed in area of high land value. Intercropping and machine operated intercultural operations are very difficult in this system of planting.

Advantages

1. 15 per cent more plants can be accommodated as compared to square system.
2. Six plants are planted at a corner of the hexagonal and 7th plant in the centre is called septule.

Disadvantages

1. Lay out is difficult and cumbersome.
2. Intercropping is very difficult.

v) Contour System

This system of planting is followed in the hilly area or high slop area. It becomes very difficult to level the area so the terraces are formed for planting the trees. Terraces are made when the slop is more than 10 per cent. Terraces are formed to irrigate easily and avoid soil erosion. In this system square or rectangle system of planting is followed to plant the trees. Layout is done by making the lowest line as a base line thereafter planting towards top.

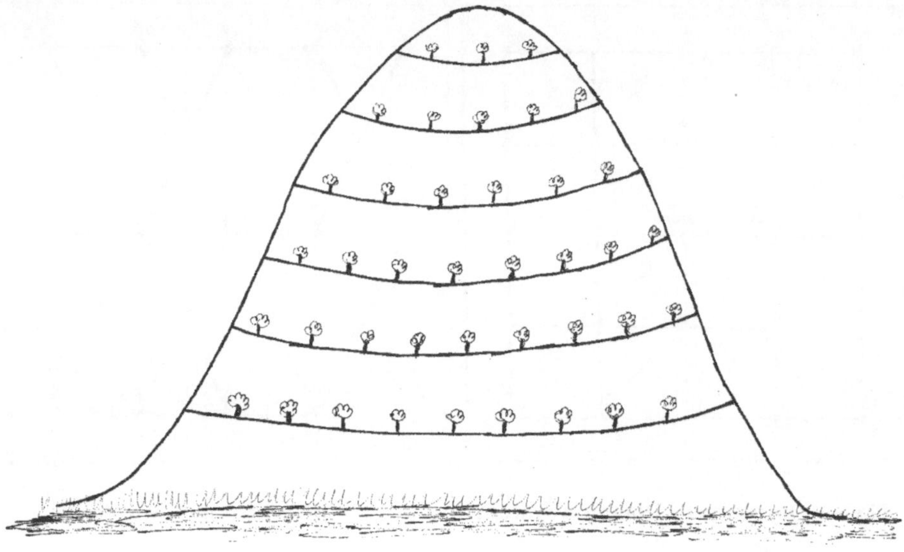

Fig 6.6: Contour System

vi) Single Hedge Row Planting System

In this method of planting each row of plants is planted at recommended space. The space between plants within row is usually one half to one third of the distance between rows. This method is used for high density planting.

Fig 6.7: Single Hedge Row Planting

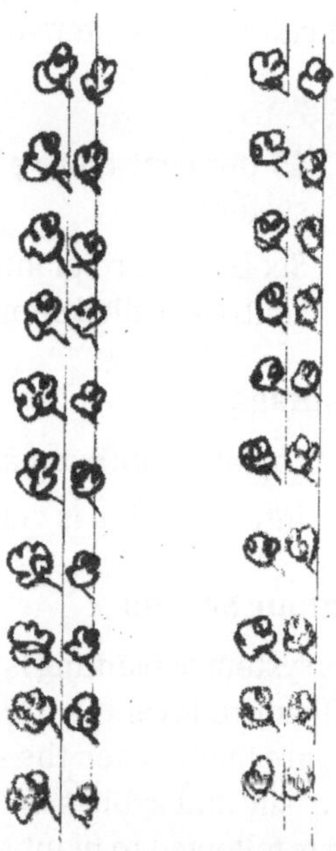

Fig 6.8: Double Hedge Row Planting

vii) Double Hedge Row Planting System

In this method pair of row is planted at close spacing and kept a wide gap between two pair of rows. It accommodates about 22 per cent more plants than single hedge row system. It is practiced for high density planting.

Block plantation: In this method different blocks are prepared in the orchard for plantation and space for monitoring or other activities is kept between block.

Planting Distance

The minimum vertical distance between any two plants is called planting distance and it depends upon various factors. The principles followed to decide the planting distances are as under:

1. Trees are fully grown when branches do not interlock but fringes of trees touch each other.
2. The root area of the tree will spread over the top area so proper room is available for the roots to feed nutrient and water without competition.

Factors which decide the planting distance are the following.

1. **Kind of fruit trees**: Mangoes are planted at a distance of 10 m x 10 m, guavas at a distance of 6 m x 6 m while papayas are planted at a distance of 2 m x 2 m.
2. **Rainfall**: Less spacing should be given in low rainfall areas than the high rainfall areas because less moisture reduces plant growth/canopy which reduces evaporation losses.
3. **Soil type and soil fertility**: In heavy soils the top and root growth are limited while in fertile soils plant growth will be more. So less spacing should be given.
4. **Rootstocks**: The growth of scion varies with the rootstocks. The same scion of different rootstocks results in varied canopy size so require different planting distances for different rootstocks *e.g.* apple.
5. **Pruning and training**: The pruning and training systems require different spacing. Trees trained on head system need less spacing than the other training system.
6. **Irrigation system:** Plant growth in flood irrigation will be more while less.

Planting Density

Optimum number of plants should be accommodated per unit area for good yield and quality of fruits. In general, if the spacing is too wide, it is obvious that the yield per unit area would be greatly reduced. Only in very exceptional cases this would be justifiable. Planting trees at closer spacing are more profitable if sufficient amount of food materials and nutrients are applied. But overcrowded or too close spaced plants grow tall and it becomes difficult for pruning, spraying and harvesting. There is competition for sun light, rooting area and nutrition. The trees give less yield and produce smaller fruits of poor colour. Intercultural operation also becomes difficult in the closely planted orchards. Close planting results in a greater yield per unit area during early stage of the tree but increase competition in later years which decrease quantity and quality of fruits.

Low density planting or traditional planting (100-250 trees/ha)

- ✰ Each mature tree gets ample space and illumination.
- ✰ Less pruning followed.
- ✰ Dwarfing rootstock not used.

Medium density (250-500 trees/ha)

- ✰ Spacing reduced.
- ✰ Tree require heavier pruning.
- ✰ Moderate dwarfing rootstock used.
- ✰ More labour required in training and pruning.

High density planting (500-1000 trees/ha)

Planting more number of plants per unit area to harvest maximum sun light and utilization of soil, water, nutrients more efficiently.

High Density Planting System

Accommodate more number of plants per unit area by planting trees at a closer spacing than the recommended one, using certain special techniques with the objective to obtain maximum productivity per unit area without sacrificing quality is called as 'High density planting' or HDP. This technique was first established in apple in Europe during sixties and now majority of the apple orchards in Europe, America, Australia and New Zealand are grown under high density planting. In this system planting densities recognized for apples are low HDP (< 250 trees/ha), moderate HDP (250-500 tree/ha), high HDP (500 to 1250 trees/ha) and ultra-high HDP (>1250 trees/ha). Recently, super high-density planting system has also been established in apple orchards with a plant population 20,000 per ha. In some orchards, still closer, planting of apple trees is followed (say 70,000 trees/ha), which is often referred as **'meadow orchards'**.

Advantages of HDP are:

(i) Early cropping and higher yields for a long time; the average yield in apple is about 5.0 t/ha under normal system of planting and it is about 140.0 t/ha under HDP.

(ii) It reduces labour costs.

(III) Improve fruit quality.

(iv) Effective utilization of resources.

High Density Planting

- ☆ Tree size controlled using dwarfing root stock.
- ☆ Chemical used daminozide (SADH or alar), paclobutrazol (cultar), maleic hydrazide (MH) and ethephon.

  e.g. Number of plants per hectare under high density planting:

  Banana cv. Robusta (1 m x 1 m=10000 plants),

  Papaya cv. Pusa Nanha (1.25 m x 1.25 m =6400 plants),

  Pineapple cv. Kew (0.25 m x 0.35 m x 0.75 m= 72727 plants),

  Mango cv. Amarpali (2.5 m x2.5 m=1600 plants),

  Litchi cv. Shahi (2.0 m x1.5 m=3333 plants).

Methods of Controlling Tree Size

1. **Genetic control:** Dwarfing rootstock, scion and inter stock.

   Dwarfing rootstocks: Dwarfing rootstock of Mango: Vellaikolumban, Kurukkan, Totapuri Red Small, Olur, Moovandan and Creeper Mango.

   Apple: M_9, M_{27}

   Guava: Aneuploid No. 82

   Citrus: Troyer Citrange and Karna Khatta
2. **Mechanical control**: Hedging and topping (remove apical dominance), pruning etc.
3. **Chemical control**: Daminozide/Alar/SADH @ 1000-2000 ppm

   Chlormequat/CCC/Cycocel @2000-3000 ppm

   Paclobutrazol/Cultar @ 1000-2000 ppm

   S3307/uniconazole @ 100-150ppm

   Ethereal/ethaphon @ 300-500 ppm

   TIBA @ 50-100 ppm

   Maleic Hydrazide (MH) @1000-1500 ppm
4. **Interplant competition:** Competition for water and nutrients also reduces the canopy size.
5. **Cultural practices:** Regulate irrigation, nitrogen supply, root restriction, spacing arrangement.
6. **Viral dwarfing:** Strain of exocortis virus which have dwarfing influence have been used in Australia.

Planting and after Care

Layout of an orchard: First of all, set the stakes on the base line at an appropriate distance. Then fix the sticks in whole field before planting.

Preparation of land: Levelling, ploughing and manuring.

Digging pit: Digging of pit of 1 m x 1 m x 1 m dimension and keep it open for 1-2 month.

Filling pit: Make a mixture of 50 per cent top soil with 50 per cent FYM or organic manure to fill the pit. It should be filled at least 1 month before planting. Apply irrigation after filling of pits to settle down soil mixture.

Planting: Plant with or without earth ball taking care of no air pocket left at the base while planting.

Planting seasons: Rainy season (July-August) and January-February (bare rooted in deciduous plants).

Early care: Removal of sprouts from stocks, protection from hot and cold weather (white wash/cover with paper or polythene/paddy straw), protection from insect and pest. In case of any damage apply $CuSO_4$ mixed in white wash on the stem @3 g/litre.

Irrigation: Preparation of basins for frequent irrigation.

Manuring: Once in a year, about 2-3 months before flowering and it must be followed by irrigation.

Stacking: To keep the trees erect growing and avoid falling of the tree.

Gap filling: Replace the plant died after transplanting.

Chapter 7

Principles of Training and Pruning

Training

It is judicious removal of the tree branches to give proper shape, increase quality and production of fruits.

OR

Techniques in which plant part is removed to control shape of plant, volume or size of canopy and direction of plant growth.

OR

Orientation of plant in space through techniques like tying, staking, propping (bamboo crutches to protect from bending or falling), spreading, supporting over a trellis or pergola (series of arches joined) in a certain fashion or pruning of some parts.

Training is related to shape and size of plants.

Objectives of Training

1. To increase the availability of sun light and air to the foliage of plant for development of maximum fruits of superior quality.
2. To make the better appearance/shape of the tree.
3. It is used to decrease the cost of inter culture operations including weeding/cultivation, spraying for plant protection and harvesting at lowest cost.
4. To develop balance between vegetative and reproductive growth of trees.

5. To remove water sprouts.
6. To develop strong framework of tree.

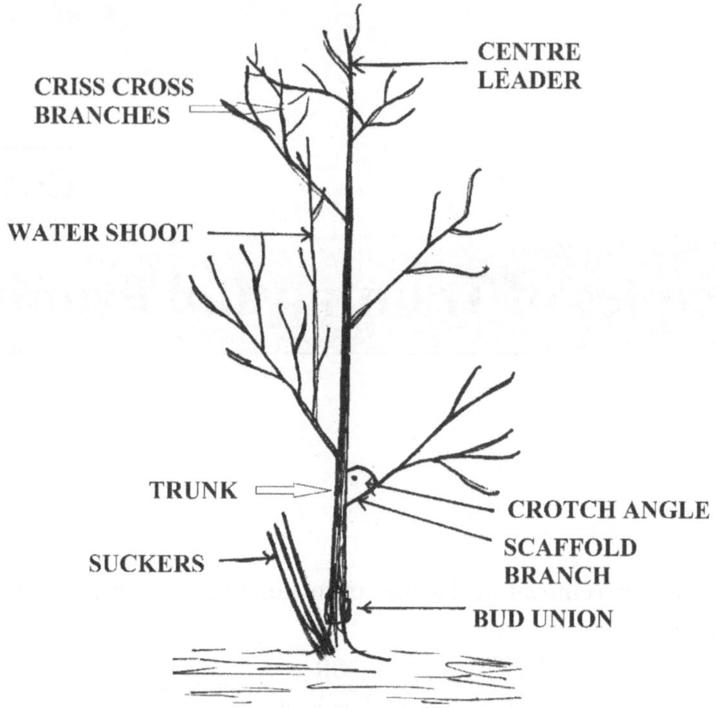

Fig 7.1: Sketch of Fruit Tree

Principles of Training

- ☆ Training should be started from time of planting.
- ☆ Fruit trees should be trained through single stem system up to certain height. However, fruits like pomegranate, fig and custard apple are trained through multi stemmed training system.
- ☆ Remove the apical dominance to facilitate the emergence of side shoots and retain better shoots after training.
- ☆ The shoots having narrow crotch angle are discarded.
- ☆ Water sprout, drooping branches should be removed regularly.

Methods of Training

Training methods decided on the basis of growth of the tree, purpose of tree (yield or aesthetic value) and climatic conditions, purpose of growing, planting method followed, requirement of mechanization *etc.* therefore, it is necessary to have complete knowledge.

Training of Woody Perennials

Woody perennial plants required strong frame work for a longer period as these have long life. The training method of these trees depends on the purpose of planting.

(i) Open Centre System (Vase shaped)

In open centre system the main stem is allowed to grow up to 75 cm after that main stem is headed back to make open center and encourage the lateral scaffold branches. There is better interception of light by all the shoots of the tree and all the branches are capable of bearing flowers and fruits. This is common in peaches, plum and apricots.

Fig 7.2: Open Centre System (Vase shaped)

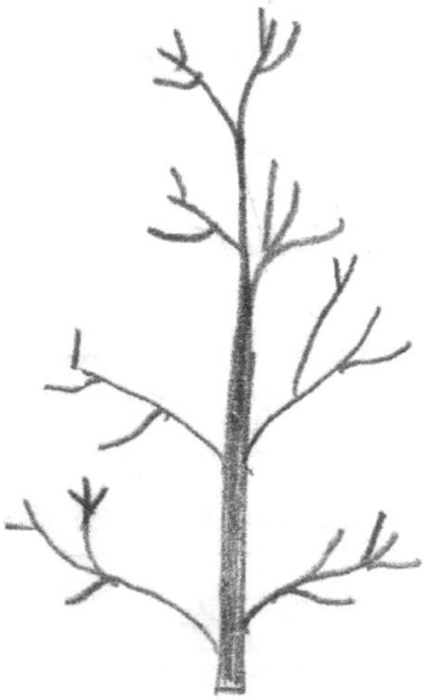

Fig 7.3: Central Leader System (Closed centre)

(ii) Central Leader System (Closed centre)

In many unpruned deciduous trees, there is a central branch growing upward with several branches growing out at different heights. This is the best method the bear the crop load and survive against strong wind. In this training system the main branch of the plant is allowed to grow permitting side branches. The first branch is allowed to grow at 40-50 cm height from the ground level and other branches are allowed to grow on main stem at a distance of 15-20 cm.

Due to less interception of light by lower branches, they remain unproductive and fruit may not be able to develop proper colour. The bearing is confined to the top outer portion of the trees, because the nature of distribution of branches. Hence, it is called central leader system and commonly in used in apple, pear, mango and sapota.

(iii) Modified Leader System

This system is most acceptable for commercial fruit cultivation. The upper most branch is not allowed to grow vertically. This is achieved by heading back of center leader after it has made some growth or after 4-5 years of planting.

Fig 7.4: Modified Leader System

The plant trained through this system attained moderate height. All retained branches receive ample sun light and there is better production on the tree. This is very suitable method and practiced in almost all the regions. In this method moderate pruning is required. The branches are well distributed to harvest maximum sun light in the interior canopy of the tree. Due to controlled/limited height spraying, harvesting and pruning are easier. It is common in apple, pear, cherry, plum and guava.

Some Special Methods of Training

(i) Bush System

Fruit trees grown in this system resemble to bush and tree height is controlled by a dwarfing scion or dwarfing root stocks. It is best suited for high density planting. Modified leader system or any other system of training can be used. In this system, the height of the plant is kept 2 meters. During first year the plant is headed back at a height of 70 cm. No shoot is allowed to grow up to a height of 25-30 cm above which 3-4 branches are allowed. The plant acquires a shape of bush *e.g.* apple.

(ii) Pyramid System

In this system, the plants are trained in a fashion so that the lower branches may remain longer and upper branches gradually smaller. The plants are pruned from the tip of the main stem and branches to maintain pyramid shape (3000 plants/ha).

(iii) Espalier System

The word espalier is of French origin means fence or fruit wall. The tree trained to a system consists of three to six tiers of horizontal branches grow one foot apart from another at right angles to the main stem.

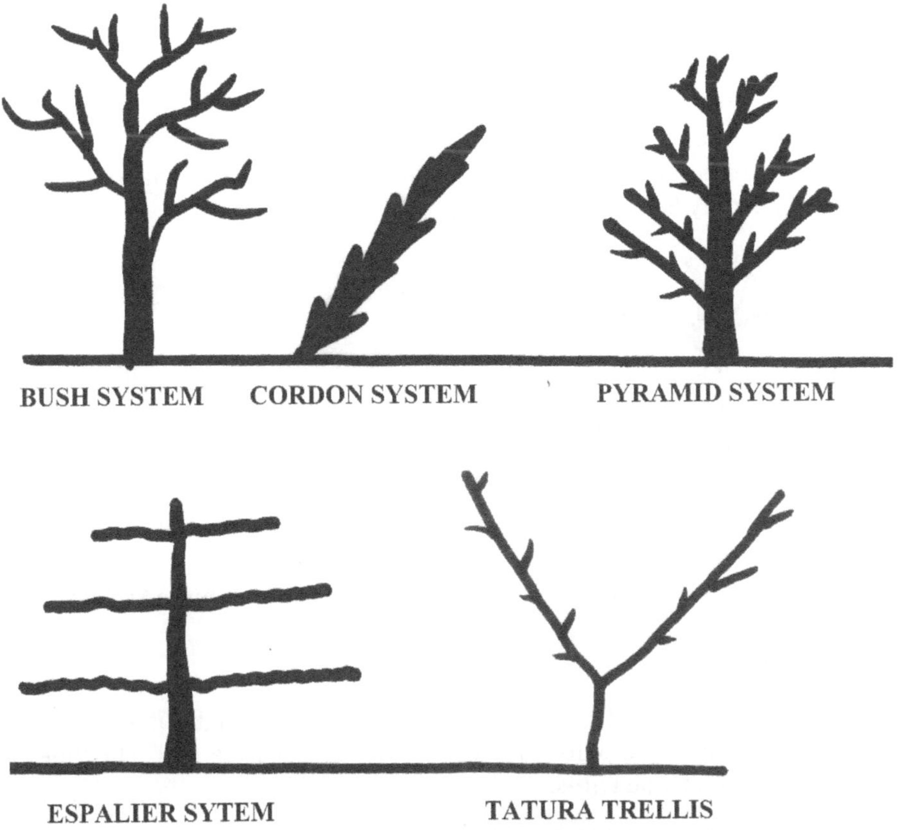

Fig 7.5: Special Methods of Training

(iv) Cordon System

Cordon refers to closely spurred single stemmed tied to a support with the help of wires or bamboo canes either in vertical or horizontal position. In this system tree limbs are allowed to grow in opposite direction with the help of wires or frames in the early years. Depending on the number of main stems trained along with wire, the system is known as single cordon, double cordon and triple cordon in crops like grape and passion fruit.

Control of tree vigour is very important. If the tree is too vigorous fruit set will be very low and the tree may produce many water shoots (2500 plants/ha).

(v) Tatura Trellis

In this system two scaffold branches are allowed at 60° in a configuration of V shape. This is a system of training fruit trees along with trellis of wire to harvest early and high yield without use of dwarfing root stock. This system is developed in Tatura, Australia for planting high density peach.

Training Methods for Grape Vine

i. Head System

Plants trained through this system develop in a bush shape. During early years the vine requires support. The plant is allowed to grow to a height of 75-90 cm. After 3-4 years, main stem become strong and vine take the shape of bush *e.g.* Perlette and Beauty Seedless.

Fig 7.6: Head System

Fig 7.7: Kniffin System

ii. Kniffin System

In this system two rows of wires are stretched at a height of 1.05 and 1.65 meters from ground level with the help of iron or concrete poll. The vines are allowed to grow on these wires.

iii. Telephone System/Overhead Trellis System

This system is known as six canes system. The polls are erected at a distance of 3.6 to 4.8 m. At the terminal end of poll there is 1.2 m long arm, wires on this terminal end arm strung about 30-45 cm.

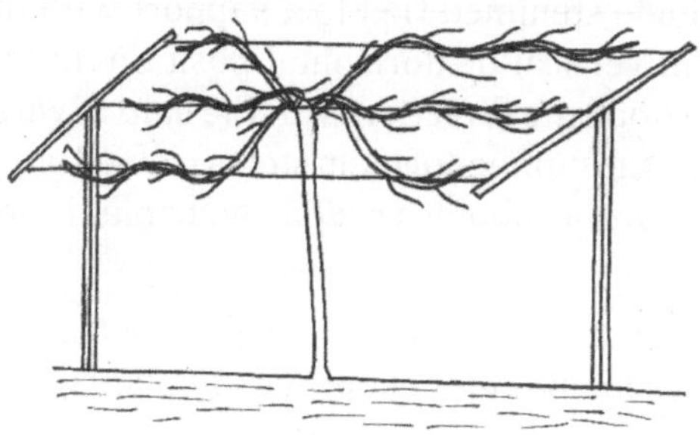

Fig 7.8: Telephone System

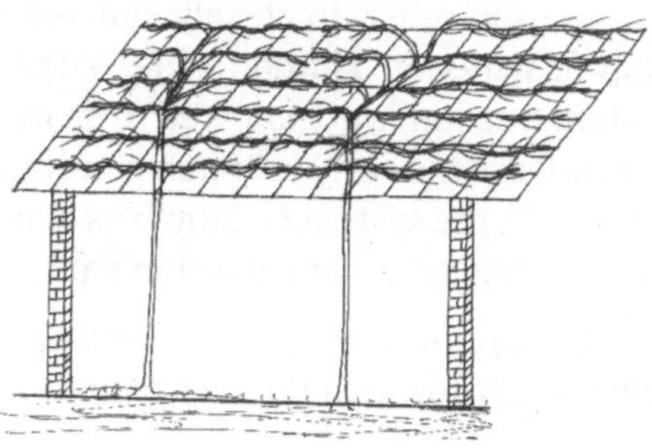

Fig 7.9: Bower System

iv. Bower System

Network of wire is created and fixed 2.0 to 2.4 m above ground level at 45 cm apart on poles to develop a roof like structure. This is also called as Arbour or Pergola. Planting distance 3 m x 3 m.

Precautions during Training

- ✰ Handle newly planted trees carefully.
- ✰ Select suitable and strong stakes only for sprouting.
- ✰ Avoid any unwanted branches on main trunk.
- ✰ Treat wound with fungicide.
- ✰ Removal of unnecessarily scaffold branches.

Pruning

Pruning refers to removal of excess or undesirable branches, shoot, root *etc.,* to improve the shape, to influence its growth, flowering and fruitfulness, to improve quality of the product, or to repair injury.

Objectives

- ✰ Remove the unproductive shoots, water shoots, suckers; diseased, damaged, broken and crisscross branches.
- ✰ Increase aeration and penetration in the interior part of the tree canopy.
- ✰ It improves quality and distribution of the fruit on the canopy of the tree and make a balance between vegetative and reproductive growth of the tree.
- ✰ Effective cultural operations like application of pesticides and harvesting.

Types of Pruning

i. **Heading back:** Removal of terminal part of shoot/stem but their basal portion is not removed. It allows/stimulates the development of lateral branches/shoots.

ii. **Thinning out:** Entire removal of unwanted branches from the base or attachment.

iii. **Spur pruning:** Removing of spurs. It results in stimulation of vegetative growth by the reduction of the fruiting wood.

iv. **Fine pruning:** Thin or pruning of weak branches.

v. **Disbudding**: It consists of removal of floral or vegetative buds before they resume growth.

vi. **Dehorning:** Removal of wood after 7-10 cm thick stubs all over tree.

vii. **Ringing or girdling:** Removal of bark or phloem portion of stem of about 3 cm length.

Time of Pruning

- When the tree has gone into complete rest/dormancy and it should be completed before the bud starts swelling/initiation.
- It is regular feature in some species to make the balance between the vegetative and reproductive growth. It is also helpful for regular bearing and quality fruiting.

Factors to achieve successful pruning

- It should be completed before the formation of vegetative buds.
- Age of the pruning wood must be kept in mind for more reproductive/ fruiting buds.
- The person/labour/owner involved in training should have complete knowledge about the bearing habit of the trees for its proper growth and fruiting in the ensuing season.
- In consideration of these factors we should have complete knowledge about bearing habit of the tree/plant.

Effects of Pruning

During pruning above ground vegetative part is removed to make the balance between root : shoot and vegetative: reproductive growth.

Effect on Growth

- The pruned tree tries to make balance between roots and shoot growth. However, total growth attained by an unpruned tree is always greater than that of a pruned tree. The growth induced in response to pruning does not compensate the portion removed by pruning plus growth which would normally occur on that point.
- Trunk and branch of unpruned trees show a greater thickness than the pruned trees.
- Ultimately pruning reduces total growth of the tree and resulted in control the size of tree.

Effect on Flower Bud Formation

Pruning develop flower bud development in young trees which have not yet flowered. The amount of regrowth that follows pruning is at the expenses of flower bud formation. As pruning severity increases the flower bud production decreases proportionately. Excessive pruning in bearing tree can result in excessive vegetative vigour, reduction in spur and flower development and decrease in yield.

Effect on Fruit Setting

Pruning increases percentage of flowers which develop into fruits. The elimination of certain growing points indirectly increases the supply of water and nutrients to remaining parts.

Effect on Yield

Due to reduced flower production, there is reduction in total yield in pruned trees. A light and moderate pruning tends to increase the marketable yield.

Effect on Size, Colour and Quality of Fruits

Increase in leaf surface per fruit increased fruit size. Reduced number of fruits on pruned trees also results in increased size, colour and quality of fruits. More light penetration also results in increased colour development.

Effect on Tree Longevity

Unpruned trees bear heavily as compared to pruned trees. So pruned trees live longer.

Rejuvenation

Heavy pruning done to old, weak, less yielding orchard tree done induce fresh growth is termed as rejuvenation. Such pruning brings about invigoration (to give energy) but does not return the tree to juvenile phase. Plant breeder are interested to shorten the period from seedling to first flower and often refer to this entire period as juvenile phase. The retention of juvenility is important for propagation because plant roots better from juvenile cuttings or explants as compared to adult tissue of the same plant.

- ☆ It is a technique or method of rejuvenation with the objective to upgrade seedling of inferior varieties to superior commercial cultivars or the desired variety which is suitable for domestic as well as export market.
- ☆ The technique involves grafting/budding of desired variety on scion shoots emerged after pruning or rejuvenation by adopting different methods of grafting during suitable season (Season of top working varies from species to species, as it depends on availability of good scion on mother plant/desired plant). The thickness of scion shoots depends on the method used for propagation. In general, both scion and stock should have same thickness.
- ☆ Top working practiced successfully and recommended in different crops like mango, aonla, sapota, guava, cashew, tamarind, jackfruit *etc.*

Advantages of Top Working

1. Increase productivity of tree.
2. It is used to convert inferior seedling/variety to superior one with a cultivar of desirable traits.
3. Many varieties can be grafted on single plant. Pollinizer can also be grafted on the same tree for better pollination and increased yield.
4. Income during rejuvenation year can be compensated by selling of pruned wood.

Disadvantages

1. The tree may die due to excessive pruning and pruning done by untrained person.
2. Plant should be managed properly after pruning for proper growth and fruiting.
3. Rejuvenated trees will not give full yield for initial 2-3 years.
4. The chances of insect pest and diseases are more in newly rejuvenated tree.
5. Skilled labour is required for thinning of shoots, removal of side shoots *etc.*
6. Need skilled person for budding and grafting.

Precautions

- ☆ Pruning should be done when plant is dormant.
- ☆ It should be done with sharp edged secateurs.
- ☆ Apply Blitox/Bordeaux mixture at the cut part to avoid fungal infection.
- ☆ Avoid splitting of branches while pruning.

Shoot: The new leafy growth of current season.

Cane: The ripened shoot of past season.

Lateral: The side branch of a shoot or cane.

Spur: The basal portion of cane left after pruning.

Fruit spur: The spur intended to bear fruit.

Renewal spur: The cane cut back to one or two buds to produce fruiting wood for the next year.

Trunk: The main stem of the plant.

Chapter 8

Juvenility & Physiology of Flowering

Juvenility

It can be defined as the physiological state of a plant during which it cannot be ready to induce flower. This state is followed by transition phase, in which flowering occur.

Depending upon the environmental and genetic factors, juvenile phase of woody plants may be very short or very long. It can be shortened by increasing the growth rate of young seedlings, because minimum juvenile phase must be attained to reach the adult stage.

The juvenile tissue remains in situ at the base of the tree for its entire life. Thus, one cut off on 80-year-old seedling tree at the lower trunk, the new growth from latent buds at the base would be juvenile until it once again grew into the adult phase. In contrast to seedling plant, commercial cultivar budded on seedling stocks are entirely adult above the bud union.

- ☆ Plants are non-reproductive for several years, even if given correct environmental conditions. Longer juvenile period generally results in a larger plant.
- ☆ Length varies with species (1-45 years).

Also, may have different characteristics:

a. Leaf shape (lobed vs. not lobed)
b. Ease of rooting
c. Stem anatomy (cambial growth)
d. Presence of thorns (*e.g.* honey locust)
e. Production of pigments

Rejuvenation

(Covered in Chapter 7)

Physiology of Flowering

Transition zone: Zone between juvenile and adult tissue.

Flowers are reproductive structure of angiosperms. Flower formation is a transition phase from vegetative to floral state in the life cycle of a plant. The shoot meristem is reduced to develop sepals, petals, stamens, carpals in place of leaves. The plant must attain specific state of 'ripeness to response' before it flowers. Once the stage is reached, then it can induce to flower.

Physiology of flowering involves the changes in the characters of cells proliferating in the meristem tissues of the shoot owing to the specific gene action and change in the phyto-hormones level. The flowering linked to the diverse environmental conditions. The controlling factors may involve hormonal balance, availability of nutrients, especially carbohydrates and interaction of these.

Vegetative (juvenile) –transition (become reproductive competent) -reproductive- plant maturation.

Flowering Physiology

Plant start flowing after completion of vegetative phase.

Process until plants come to flower is classified into three stages.

1. Differentiation of flower buds
2. Development of flower buds
3. Opening of flower

Initiation of flower bud is the conversion of plant from vegetative phase to the reproductive phase is called flowering.

Flowering physiology is to clarify the mechanism involved in switching the mode of growth from vegetative to reproductive.

Flowering occurs when flowering genes are activated. The activation of the flowering genes occurs autonomously or is trigged by environmental factors. Autonomously flowering occurs when plants reaches a certain age. The environmental factors trigging the activation of flowering gene are photoperiod, low temperature and various stresses. Photoperiodic flowering regulated by photoperiod and vernalization regulated by low temperature have been well studied.

Physiology of Flowering in Horticultural Crops

Events in the Bud Leading to Flowering

There are three process occur during the transition of bud to flowering *i.e.* induction, evocation and initiation.

i) Induction: The step of flowering at which the flowering stimulus is generated is called floral induction. Induction and induced state are defined as a physiological condition initiated in tissue by external influences such as water stress, chilling temperature or photoperiod, progressively towards floral expression even under subsequent non-initiating conditions. The event is translated to production of a putative floral stimulus or alternative in the ratio of florigenic and antiflorigenic components which may be translated to target cells in meristems.

ii) Evocations: The step at which the apical meristem received the flowering stimulus, and is irreversibly committed to form flower bud promordia is called floral avocation, these processes lead to the formation of floral primordia *e.g.* if a vegetative shoot (scion) is grafted.

iii) Initiation: In this process the evoked bud becomes recognizable flower bud and thus committed to reproductive development. This evident the broadening and flattening of the growing point with concurrently developing lobes.

Differentiation of the Growing Points

The transition from vegetative to reproductive development is marked by an increase in the frequency of cell division within the central zone of the shoot apical meristems. This leads to increase in the size of meristem and the shoot meristem is induced to develop sepals, petals, stamens and carpels in place of leaves. The first detectable change in the growing points of the bud after induction of flowering is increased synthesis of DNA and RNA.

The entire process of flowers and differentiation may take 54-112 days depending on the species.

Changes Occurring in Target Tissues

a. Biochemical Changes

i. Development of chloroplast inhibitors.
ii. Moving out of floral stimulus from leaves.
iii. Increase in total RNA synthesis and chromatins dependent RNA synthesis.
iv. DNA synthesis increases.
v. Mitotic activities increase.
vi. Number of mitochondria increase.
vii. Starch content decrease.

b. Metabolic Changes

i. Meristem height increases.
ii. Cell size increases.
iii. Lengthening of mitotic phase (second mitotic peak).
iv. Entry of cell from photosynthetic to mitotic phase.
v. Flower bud initiation.

Factors Affecting Flowering

1. Photoperiodism

The induction of flowering in response to the relative length of daily light and dark period is called photoperiodism. It is physiological response of plant in relation to daily length of dark or light period. The proportion or length of dark/light is helpful for flower induction in the plants or transforming the vegetative bud into floral bud.

The photoperiodic response is most important in the initiation of flowering. The photoperiodic signal is first received by phytochrome in the leaf and the signal from phytochrome starts the biological clock. After the biological clock measures a certain period of time (inductive photoperiod), the production of flowering stimulus in the leaf starts. The flowering stimulus is transmitted from leaf to shoot apex leading to change in the growth mode of the meristem from vegetative to reproductive. The apical meristem produces primordia of floral organs *viz.*, sepals, petals, stamens and carpels, thus generating a flower bud.

Classification on the Basis of Photoperiodic Response

i) Short day plant (SDP): The plant flowers during the shorter day length than certain critical maximum. Most of tropical plants are short day plants *e.g.* pineapple, strawberry *etc.*

ii) Long day plants (LDP): These plants initiate flowers in response to longer day length than certain critical minimum *i.e.* 12-14 hours. Plants continue their vegetative growth in day length below the critical period. Most of the temperate zone fruits are long day plants *e.g.* banana, apple.

iii) Day neutral plants (DNP): These plants flower after completion of vegetative growth instead of the photoperiod. Flowering in these plants is not affected by day length and it occurs throughout the year *e.g.* papaya, guava.

iv) Long short-day plants (LSDP): Certain plants like *Cestrum nocturnum* (night jasmine) have the dual photoperiodic requirement for flowering. Such plants need to be exposed to long day for some period followed by exposure to short day. If sequence is reversed and short days given first, followed by long days these plants fail to flower.

v) Short long day plants (SLDP): Some plants like *Coreopsis grandiflora* also require dual photoperiod for flowering such as short days first and then followed by long days.

vi) Qualitative and quantitative photoperiod response: Both short day and long day plants can further be classified into two types: Qualitative plants, Quantitative plants.

Short day plants like *Xanthium* can flower only when exposed to short day. Such plants fail to flower and continue to grow vegetatively almost indefinitely when kept under long days. These are called qualitative or obligate short-day plants.

On the other hand, plants of *Cosmos* and *Zinnia* can flower under both short day and long days but flowering is early under short day and is delayed under long days. Such plants are quantitative or facultative SDP.

Similar in the case of long day plants.

Qualitative LDP: Flowers only under long day – *Hyoscyamus.*

Quantitative: Flowering promoted by long day but can flower under inappropriate photoperiod. LDP – *e.g.* sunflower, petunia, salvia.

Critical photoperiod: It refers to the limiting photoperiod (up to which SDPs flowers) in case of LDPs the critical photoperiod is the limit above which these plants flower. Critical photoperiod varies from spp. to spp. Same photoperiod can be short day for one species but long day for other.

Photoperiod is said to be inductive if it causes flowering in plant and it is non-inductive if it does not lead to flower in the plant.

Importance of light and dark period: Interruption in dark period for short day plant and light for long day plant resulted in failure of flowering.

Accumulation of Pfr inhibit flowering in SDP and triggers flowering in LDP.

Xanthium (SDP) flower when exposed to 15 h or < light

when exposed to 9 h or > dark

In SDP – If photosynthates supplied in the form of sucrose, it flowers even in total dark of 24 h. So for SDP dark period is important for flowering.

If dark period is interrupted even by a brief flash, SDP fails to flower.

In LDP - Light period is important. Dark period can be interrupted or completely eliminated.

					SDP	*LDP*
9L	////////////////////15D///////////////////////////				Flower	Veg
9L	/////////////////////	WL	/////////////////////		Veg	Flower
9L	/////////////////////	Red	/////////////////////		Veg	Flower
9L	/////////////////////	FR	/////////////////////		Flower	Veg
9L	/////////////////////	R/FR	/////////////////////		Flower	Veg
9L	/////////////////////	FR/R	/////////////////////		Veg	Flower

L: Light, D: Dark | Red 660nm, FR 730nm

Fig 8.1: Response of Flowering to Lights

Light quality – Red light (660nm) is most effective part of spectrum in preventing flowering in SDP.

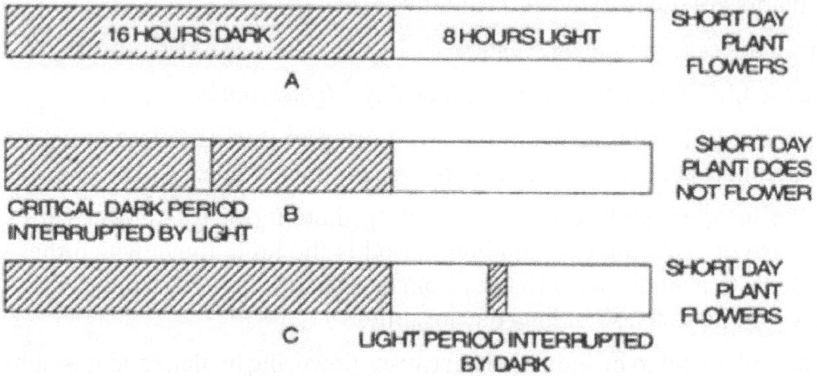

Effect of brief exposure of red light during dark and interruption of light period by dark on flowering in a short day plants.

Fig 8.2: Response of Flowering to Lights (R/FR)

Importance of Photoperiodism

1. Yield can be increased by knowing favourable photoperiods in plants.
2. Plants like radish, carrot *etc.*, can be made to remain vegetative for longer period.
3. Annuals can be grown twice or thrice a year.
4. Prevention of winter dormancy and autumn leaf fall.
5. Increased stolen formation through long day in strawberry.

2. Phytochrome Theory

Plant contains chlorophyll, carotenoid and anthocynin pigments. None of these pigments is responsible for absorbing red and far red radiations. A new pigment was discovered by Borthwik and Hendricks in 1959 in USA and it was named as phytochrome (Pr and Pfr). Phytochrome is a non-photosynthetic photo-receptor pigment that exists in two forms:

Pr– Absorb maximum light in red region (660nm) of spectrum.

Pfr- Absorb maximum light in far red region (730nm) of spectrum.

These two types of pigments are inter-convertible by two methods photo-conversion and thermal conversion.

Photo conversion: When Pr form of phytochrome absorb red light, it becomes transformed into Pfr form. Similarly, when Pfr absorb far red light it becomes converted Pr form. This is known as photoconversion. It is very rapid and complete within 1-4 minutes.

Thermal conversion: Only Pfr form of phytochrome shows this conversion. Pfr converted to Pr in dark. This mode of conversion is very slow and takes few hours. It is dependent upon temperature so called as thermal conversion.

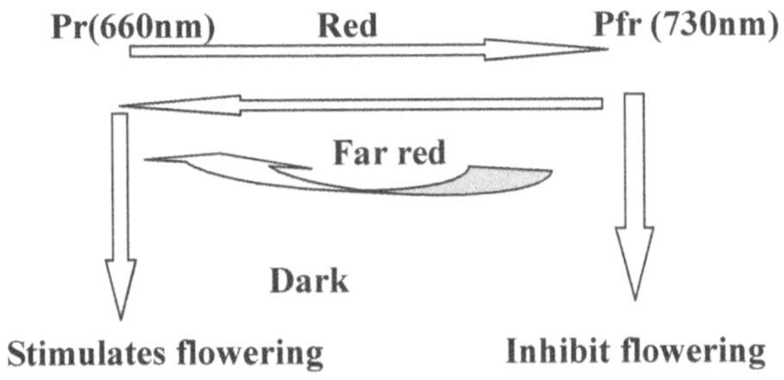

Fig. 8.3: Role of Light is Stimulation of Flavour

Pr form does not show thermal conversion.

Only Pfr form of phytochrome - physiological active, Pr is inactive.

Accumulation of Pfr - Inhibit flowering in SDP.

- Trigger flowering in LDP.

Leaves perceive the effect of short day but not shoot apex. A single leaf or even a fraction of leaf (2 cm^2) is sufficient to perceive the effect of SD and cause of flowering. Young leaves are sensitive.

It is believed that the phytochrome influence the flowering in plants.

It is that phytochrome controls flowering in plants. P_{730} promote flowering in long day plants and P_{660} promote flowering in short day plants. However, P_{730} suppresses flowering in short day plants and P_{660} suppresses in long day plants. Sun light functions as red light. At the end of light *i.e.* in evening after sun set, Pfr form pre-dominates while at the end of dark period, *i.e.* before sunrise, Pr form predominates.

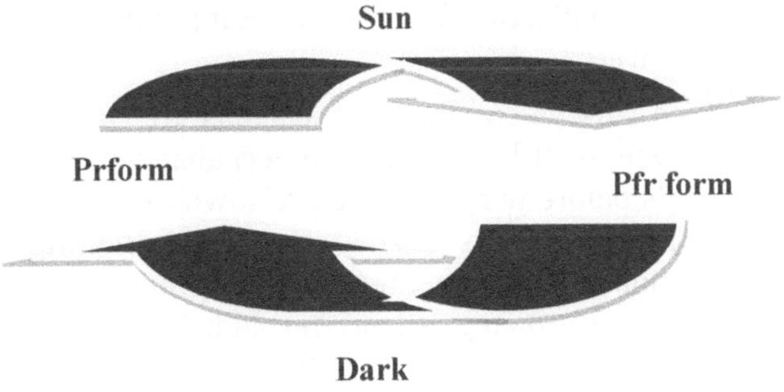

Fig 8.4: Conversion of Light

- ☆ During long night Pfr form is converted to Pr form. If the long dark period is interpreted with red light, insufficient amount of Pr form inhibit flowering in short day plants.
- ☆ The ratio of Pr/Pfr influence flowering in plants.
- ☆ High Pr/Pfr ratio favours flower formation in short day plants while low Pr/Pfr promotes flowering in long plants.
- ☆ Low Pr/Pfr inhibits flowering in short day plants while high Pr/Pfr inhibit flowering in long day plants.

3. Florigen

Site of perception of SD - flowering at shoot tip instead of leaf, some substance/ stimulates moving from leaf to buds. This hormonal substance was named as florigen by Chailakhyan,1936 (Russian). Florigen is a hypohetical flower causing hormone because it has yet not been extracted and purified. It can move from one plant to other plant through graft union.

4. Vernalization and Flowering

It refers to requirement of low temperature to seed or plant which results in promotion of flowering. Some seeds or buds also need low temperature for breaking dormancy. But low temperature requirement for processes other than flowering is not vernalization. These are environmental factors regulating flowering other than night length. Temperature is also important, some plants required exposure to low temperature 0 to 10°C for a few days to weeks for flowering. Induction of flowering by low temperature is called vernalization (vernalin discovered by Melcher, 1936). It is observed in many long day plants, but is not correlated with photoperiodism. The length of chilling may vary from species to species ranging from four days to few months and the most effective temperature is 4°C.

Qualitative vernalization (biennial forms): If plant fails to flower and remain vegetative indefinitely, without being exposed to low temperature *e.g. Hyoscyamus*

Quantitative vernalization (annual forms): On other hand some plants flower even without vernalization but the low temperature treatment hasten flowering *e.g.* pea, gram *etc.*

The plants which survive two growing seasons are called biennial plants. Its example is Hyoscyamus. It has qualitative vernalization requirement. Their first growing season is before winter. These are sown in summer. They develop vegetative body in first growing season. They store food in their roots. Then winter comes vegetative portion dies. After winter the underground root give rise to new plant body. It is their second growing season. The plant develops flowers and seeds. It is harvested in spring. These plants fail to flower without vernalization. They remain vegetative indefinitely without vernalization.

Stimulus of vernalization: Vernalization effect perceived at stem tip.

- Plant can be vernalized - if stem tip is chilled and rest of the plant kept at high temperature.
- In seeds - Embryo is the site of vernalization stimulus reception, not the endosperm.
- Vernalization is not possible in anaerobic conditions.
- It has been postulated that vernalin (Hypothetical hormone) is transmitted from vernalized plant to non-vernalized through grafting.

Significance of Vernalization

- Production of more than one crop in a year by shortening the vegetative phase.
- Protection from freezing injury by sowing winter crops in spring.
- Tropical plants can grow in temperate area (short summer).

Devernalization

Loss of suboptimal vernalization effect by subsequently exposing the seed or plant to high temperature (30-35°C). After optimum vernalization, devernalization is not possible.

Light or dark has no effect on vernalization. However, for LDP light is important for flowering after vernalization.

- Carrot (an obligate vernalization) plants remain rosette during vegetative stage. When vernalized these plants show flowering accompanied by rapid stem elongation. Such elongation of rosette stem is called bolting.
- In such rosette plants vernalization can be replaced by GA. It cause both flowering and bolting without low temperature.

Other Factors for Induction of Flowering in Fruit Crops

1. **Water relations**: The tropics have distinct rainy and dry season. Sometimes water stress substitute the chilling. Excessive rainfall or irrigation maintains extension of growth late into summer, often reduces flower bud formation, but this is inhibited under moderate drought and almost prevented under severe stress. Plant water stress has generally been presumed to provide the stimulus for flowering.
2. **Nutritional factors**: The concentration of sugars should be greater than nitrogenous compounds for flowering. Higher C/N ratio and adequate mineral elements or proper balance for normal flower development. Deficiency of nutrients like Zn, Cu and B is detrimental to flowering.
3. **Plant growth substances**:
   i. **Gibberellins**: GA is a primary inhibitor of floral initiation. A specific stage must be attained for a bud to induce/initiate flowering. Initiation

at the apex begins only after a critical node number is reached. The time interval between successive nodes is termed as plastichron.

ii. **Auxins**: Auxin is mainly used for floral induction in pineapple. The induction of flowering in pineapple with auxin was first reported as early as 1939. Among the auxin substances, NAA and NAA based compounds like planofix and celmone have been reported effective in induction of flowering @ 10-20 ppm.

iii. **Cytokinins**: It is generally seen that upward flow of cytokinins in xylem sap of trees reached a peak in spring near the time of full bloom.

iv. **Ethylene**: Smoke is known to contain ethylene, which is a flowering hormone.

4. **Other chemicals**: Use of TIBA (Triido Benzoic Acid), SADH (succinic acid-2,2-dimethylhydrazide), PBZ, CCC *etc.*

5. **Cultural practices**:

   i. **Spreading, orientation, bending and shoot type**: This is widely practiced to increase the fruit production in **guava, bending is** especially practiced in the erect growing varieties.

   ii. **Girdling**: Girdling is the removing of phloem portion, it is practiced to force the flowering in mango, which is not bearing normal crop.

   iii. **Defoliation**: It is practiced to induce the dormancy in the crop for induction of flowering.

   iv. **Notching:** The removal of small narrow bark strip above the dormant bud *e.g.* fig.

   v. **Pruning and pinching**: Pinching is required in crop which bears flower on current season growth.

6. **Horticultural Traits:** Age, crop load, root stocks, type of shoots *etc.*

Conclusions

1. The flower bud development involves the transformation of vegetative apex to a reproductive structure.
2. These events occur during transition of bud to flowering *i.e.* induction, evocation and initiation.
3. First visible sign of differentiation is when the flat apical meristem becomes domed then central meristem is partitioned and the pith meristem developed.
4. Photoperiodism along with hormonal balance is critical factors for flowering rather than C: N ratio in fruit crops. Low temperature regulates flowering in long day plants and water stress enhances flowering in mango and citrus.

5. Various theories have been put forth by many workers to explain the mechanism behind flowering but none is still holding good.
6. Flowering in fruit crops can also be influenced by various cultural practices like bending of branches, defoliation, chemical treatments, nutrient regulation, girdling and pruning.
7. Growth regulators have striking effect on flowering. While auxin promote it, GA inhibit it, NAA is being used in mango and pineapple to induce flowering. Ethephon is a major growth regulator regulating flowering in pineapple. Various growth retardants *i.e.* Alar, Cycocel, TIBA (anti auxin), SADH and PBZ stimulate flowering by inhibition GA synthesis. Paclobutrazol has been widely used in mango to induce flowering during the off year.
8. Molecular mechanisms are still not well understood, however, many genes that regulate flowering have been isolated. Hence, flowering is a complex phenomenon regulated by number of factors.

Fruit Bud Differentiation

Inflorescence

Flower head of plant which includes flower, stalk and brackets.

Types of Inflorescence

1. **Solitary flower:** Flower borne on axil of leaf at the terminus shoot or on specialized stalk *e.g.* guava, almond, peach, apricot, quince.
2. **Racemose:**
   i. Raceme: Flowers borne on the pedicle which are equal in length *e.g.* blueberry, black berry, ghooseberry, raspberry *etc.*
   ii. Catkin: Bear unisexual sessile flowers on pediculosis main axis *e.g.* mulberry, walnut, chestnut, peanut.
   iii. Corymb: The pedicel of lower flowers is longer than the upper ones and all the flowers reach the same level *e.g.* pear, *Cassia auriculata,* candy tuft.
3. **Cymose inflorescence:**
   i. Fasicle: Flowers formed on very shot peduncle with a pedicle of about equal length *e.g.* ber, plum, cherry.
   ii. Panicles: It is a branched raceme in which each branch has more than one flower *e.g.* mango, grapes, litchi, loquat.
4. **Anomalous type of inflorescence:** Some inflorescence which do not fit into conventional types. In case of apple terminal flower open first than basal one and then anthesis of individual flower proceed more or less upward to apex, it is suggested as racemose type.

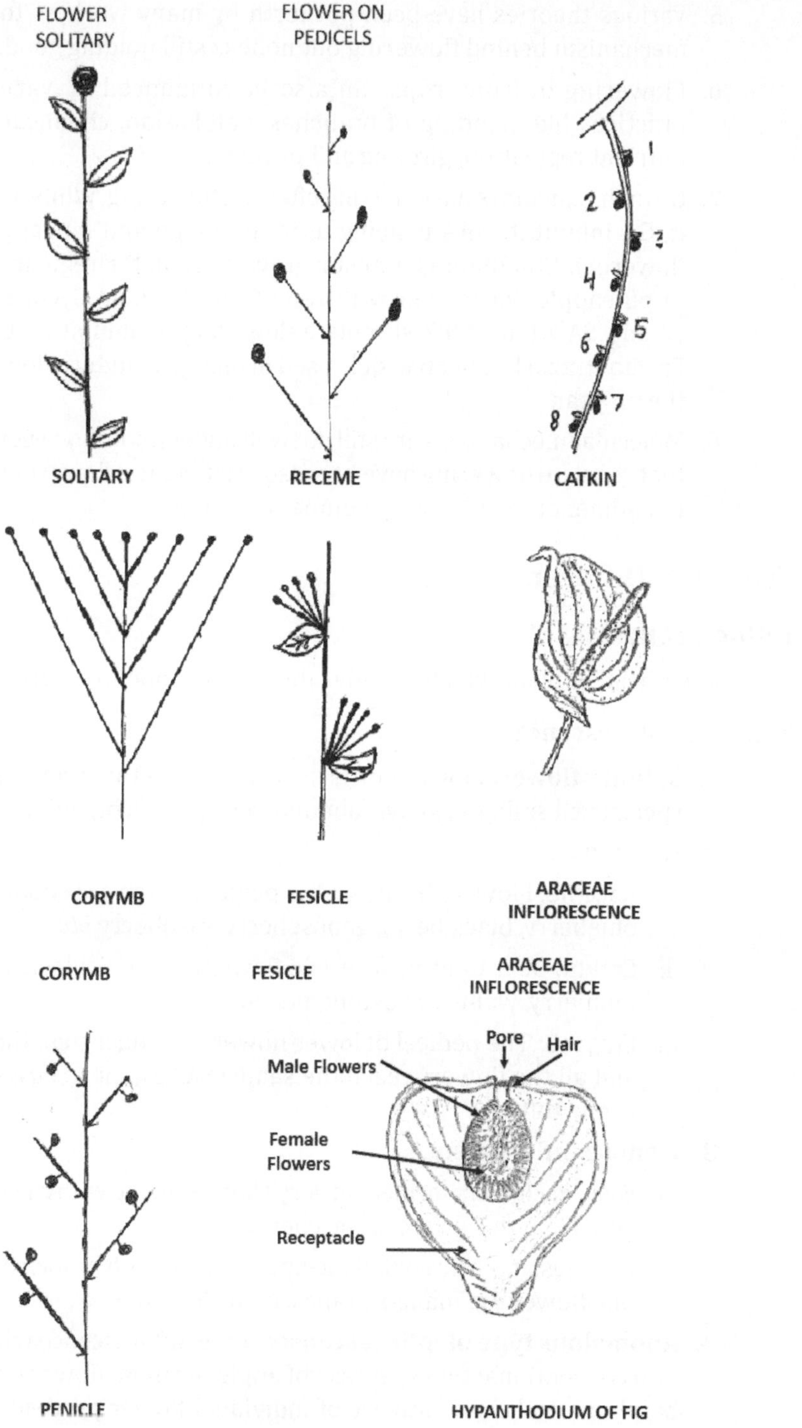

Fig 8.5: Type of Inflorescene

5. **Other**
   i. Spadix: Fleshy peduncle covered by long showy bract (boat shaped) with spike inflorescence *e.g.* banana, date palm and *Colocasia*.
   ii. Hypanthodium: It is modification of the capitulum in which the receptacle is hollowed into a flask. It opens outside the apical pore in which female flower lies towards the base and male flower lies towards apical pore *e.g.* Fig.

Bud

It is young underdeveloped shoot consisting of a short stem and a number of tender leaves arching over the growing apex.

Types of buds on the basis of position:

i) Axillary bud: Bud occurs on the axil of the leaf.
ii) Terminal bud: Bud occurs on apex of stem.

Types of bud

Simple: Give rise to flowering or vegetative shoot only depending upon the nature of bud *e.g.* cherry, citrus, coconut, date palm, peach, almond, apricot, plum, mango, loquat.

Mixed bud: Give rise to both flowering and vegetative shoot *e.g.* apple, pear, ber, grapes, guava, pomegranate, olive.

Fruit Bud Differentiation

Conversion of vegetative bud into flowering bud. The flower bud induction and differentiation process can be divided into three main stages.

1. When the stimulus trigging flower initiation reaches the apical meristem, RNA and proteins, essential for flower initiation are synthesized.
2. This is continued by the mitotic cycle when the nuclei divide.
3. By morphogenetic cycle when the flower primordial develop.

Endogenous Regulating System of Flower Bud Initiation

1. **The determining role of vegetative phase of bud development proceeding flower initiation:** Flower bud development is attained through transformation of the vegetative apex to a reproductive structure. This transformation takes place only when the structure of vegetative bud is complete. This requires that a certain number of vegetative organ primordial must already be developed on the axis of the bud.

   In Golden Delicious all the spur buds had 16 nodes by the end of July. The buds that remain vegetative did not produce any more nodes during growing season (till end of October). The buds that develop flower bud primordial produced five more nodes on the bud axils by the beginning of September. Thus, before the first visible sign of the flower initiation, there is no difference in rate of node production in buds whether they become flower buds or remain vegetative.

The treatment with growth regulators that increased bloom (daminozide) also increase number of nodes in the bud axis. In contrast, the treatment that reduced bloom (GA) produce lower node numbers.

2. **Effect of leaves**: Flower bud formation itself is a growth process and the vegetative growing shoots provide substances needed for organogenetic activity. Too weak vegetative growth lowers the number of flowers with in inflorescence but does not lower the number of flower buds. Excessive growth lowers the number of flower buds.
3. One of the pre-conditions for flower bud formation is adequate leaf area. No flower bud formation at 10 leaf/fruit ratio.
4. **Influence of hormones originating from developing seeds**: When the spurs bore seedless fruit, more than 90 per cent of spurs would flower in the next year. But when seeded fruits are developed, number of blooming spurs in the next year decreased in proportion of number of seeds present. Auxin, GA and ethylene hormones are related to flowering.

C: N Ratio

A satisfactorily general balance must exist between carbohydrates and nitrogen supply in the plants before fruit bud differentiation and initiation of flowering, which is regulated by C: N ratio. The nutritional status can directly influence the degree and quantity of flowering. On the basis on nutritional status fruits have been divided into four general groups/classes as follow:

Group I - C/NNNN

Plants in this class make poor growth and bear little or no fruits. There is a small quantity of C as compared to N in the tree. Young trees near the bearing age which have been defoliated by insects and disease or trees which are grown in shade or are over crowded may fall in this class.

Group II - CC/NNN

In this class excess nitrogen is present with sufficient amount of carbohydrates for a strong and succulent growth. Practically all the carbohydrates are used for growth and hence there is no storage. This type of balance represents the normal growth for trees and that are too young to bear or for mature trees formed into growth by severe pruning or too much manuring. This type of balance also favour more vegetative growth.

Group III - CCC/NN

A fair growth and fruits are produced. The balance of C and N is right for both growth and fruitfulness with a slight excess of carbohydrates. This is the nutritional balance sought by the producers.

Group IV - CCCC/N

This class represents very high C: N ratio. Excess amount of C and small quantity of N are present. This class represents a starved condition. Most of the old trees may found in this nutritional condition and the neglected trees also fall in this group. They often flower profusely.

- ☆ So, it is clear that as the trees come into bearing, they tend to pass from group one or second to group third and then group forth.
- ☆ The ratio indicated in the headings is not the exact ratio of C and N in plants but for the purpose of understanding, the headings represent hypothetical and give an idea of lowness or highness of the levels of carbohydrates or nitrogen in fruit plants, because nutrition is considered as one of the controlling factors for flowering.
- ☆ High ratio of carbohydrates to inorganic nutrients, particularly nitrogen (resulting in higher C/N ratio) favoured flowering and fruit set. Heavy application of nitrogen (low C/N ratio in the plant) delayed flowering in young fruit trees, while old or weak trees (heaving high C/N ratio) often flower profusely.
- ☆ The effect of nitrogen status varies with species in photoperiodicity induced plants, nitrogen deficiency often promoted flowering in LDPs while delayed it in SDPs.

Table 8.1: Time and Site of Flower Initiation

Fruit Crops	*Time*	*Flowers Born on*	*Season of Anthesis*
Apple	June - July	Terminal bud 2 years spurs	Next spring
Pear	July- August	—do—-	—do—-
Quince	Early spring	Terminal shoot current growth	Same spring
Peach	June- July	Lateral bud on one-year shoot	Next spring
Plum	July – August	Lateral buds on two-year spurs	Next spring
Strawberry	September	Crown bud	Next spring
Grape	Mid-June	Lateral bud on one-year canes	Next spring
Fig 1st crop	Late summer	Lateral bud on one-year shoot	Next spring
Fig 2nd Crop	Early summer	Lateral bud current growth	Same season
Mango	September	Terminal	Same season
Citrus	February	Current season one year old	Same season
Ber	July-August	Current season growth	Same season

Bearing Habits

It refers to the position of fruit bud with respect to vegetative growth. Based on the position of fruit bud, trees are classified as:

1. **Shoot bearing**: Bearing on shoot, may be terminal, axillary or both.

2. **Stem or branch bearing (Cauliflorous)**: Bearing directly from the main stem or branch *e.g.* jackfruit, Cacoa.
3. **Spur bearing**: Spur is modified shoot of more than one year which bears fruit bud and remains productive for a few years depending upon cultivar and species.

Fruit Development

Fruit: Fruit is the structure that arises from an ovary (or several ovaries and in some cases associated floral parts) after fertilization and support for developing seeds. It is an evolutionary adoption unique to flowering plants and is designed to aid in the dispersal of seeds by various agencies.

Stages of Fruit Development

Growth – maturation – physiological maturity – ripening – senescence.

Development: The series of process from the initiation of growth to death of plant or plant part.

Growth: The irreversible increase in physical attributes (characteristics) of a developing plants or plant part.

Maturation: Maturity is derived from Latin word *'Maturus'* means ripen. The stage of development leading to the attainment of physiological or horticultural maturity or it is the time between the final stage of growth and beginning of ripening.

Physiological maturity: Attainment of full size by fruit.

Horticultural maturity: The stage of development is optimum for utilization by consumers for a particular purpose.

Ripening: The growth and development in which transformation of fruits from an unfavourable state of firmness, texture, colour, flavor and aroma to a more favourable state of consumption.

Senescence: The last stage of development during which degradation of biological components occurs.

Fruit development: Fruit enlargement throughout the growing season is the result of cell division after flowering, cell enlargement after fertilization or both. Generally, cell division predominate in first few weeks after blossoming but later on overlap with cell enlargement phase which lasts until fruit maturity. Duration of cell division phase after anthesis is:

- 2- weeks in apricot
- 3-4 weeks in apple, peach, plum, grape
- 6-8 weeks in pear
- 4-9 weeks in orange

If cumulative increase in volume, weight or diameter of fruit is plotted against time after anthesis, the resulting curve may be:

Sigmoid: Apple, avocado, almond, date, oranges, pear, strawberry.

Double sigmoid: Apricot, sapota, seedless banana, ber, papaya, guava, cherries, peach, plum, grapes, olive, kiwi fruit, fig *etc.*

Triple sigmoid curve: Kiwi fruit

Types of curve bear no relationship to the structure of the fruit.

Kinetics of Growth – The Course of Growth (Grand period of growth) or Sigmoid Curve (Grand period curve)

Usually under favorable conditions there is a characteristic increase in the plant's growing parts. Growth is slow at first (Lag Phase), then gains speed (Log Phase) and eventually slows down (Decreasing Growth Rate) to come to a halt (Steady State). The total time during which this course of growth takes place is called as the grand period of growth. If this growth rate is plotted against time, a slanting S shaped curve is obtained which is called as sigmoid curve or grand period curve.

- ☆ The sigmoid curve represents the integrated sum of the curves for each growing organ and cell. It represents the change in size of these parts. Similarly, when dry weight is measured as an index of growth before maturity, the curve takes the well-known sigmoid form. Environmental conditions may alter growth rates but not the sigmoid form of the growth curve.
- ☆ In unicellular organisms such as *Chlamydomonas* or bacteria growth is assessed by a count of number of cells per milliliter at increasing times after that the cells are placed in a fresh nutrient medium and under environmental conditions (light, temperature *etc.*) suitable for optimal growth. Here also, there is initial lag period during which cells activate their biochemical machinery for rapid growth by synthesizing necessary enzymes. This is followed by a time period during which there is exponential increase in cell number which is called as log period.
- ☆ This period of rapid growth does not continue indefinitely and due to depleted nutrient supply, accumulation of toxic products and other limiting factors ultimately leads to decreasing cell number until the population of cells reaches a steady state in which the number of cells remains constant (stationary) or even declines. If number of cells per milliliter is plotted against time (hours), again a sigmoid curve is obtained as described earlier.

Physiological Development of Fruits

As a general principle, fruit development in terms of weight and volume tends to be sigmoidal. A period of very rapid cell division but very little increase in fruit size (stage I- logarithmic phase), is followed by a period of rapid increase in size as small, newly differentiated, dense cells develop vacuoles and assume their roles as specific tissues (stage II, Linear phase). In the final stage, as the fruit reaches physiological maturity, increase in size slows and may even stop, although biochemical changes may continue (stage III, senescence phase- growth rate decrease). There are about as many variations on this pattern as there are different types of fruit, but the sigmoidal mode is usually discernible. The orange, apple, and apricot are considered as typical examples of the development of citrus, pome and drupe fruits.

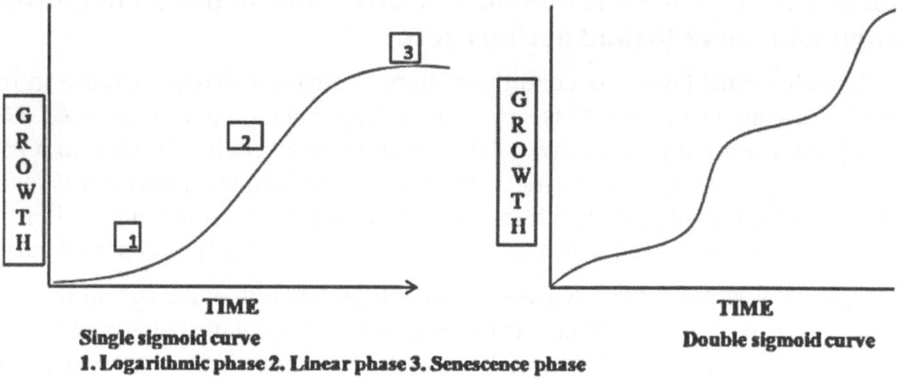

Fig 8.6: Sigmoid Curve / Fruit Development Curve

Fruit Development Stages

1. **Cell Division**
   - ☆ Predominate after bloom.
   - ☆ Smaller fruited crops generally have a shorter period of cell division. It can be extended in some fruits by blossom thinning.
   - ☆ 2 million cells in flesh of apple at anthesis require 21 doublings of cell number.
   - ☆ 40 million cells at harvest require only 4.5 doublings.
   - ☆ Most post-anthesis cell division occurs in first few weeks after bloom, but as long as 100 days.

2. **Pit Hardening (Stone fruits only)**

   Lignifications of endocarp.

3. **Cell Enlargement**
   - ☆ Predominates later in fruit development (and after pit hardening is stone fruit).

☆ Begins soon after pollination, continues through cell division stage, then at diminishing rate until harvest.

4. **Fruit Maturation**

Final weeks (days) of fruit development.

Internal Factors Affecting Fruit Growth

1. **Hormones**

If a flower is not pollinated and fertilized- the potential fruits fail to grow- it will drop (except partheocarpic fruits). It is believed that the seeds contribute a substance or substances necessary for plant development-this substance is hormonal. Because unfertilized fruits can sometimes be induced to grow by application of hormones. Auxin, GA (cell elongation) and cytokinin (cell division) will often induce parthenocarpy.

In fruits which have several seeds like kiwi fruit, strawberry, apple and grapes, the size of the fruit usually bears a direct relationship to the number of seeds present. If the distribution of seed is uneven, the fruit growth will be lopsided, with the side bearing most seeds growing more. It is suggested that the seed influence the growth of adjacent flesh, possibly by migration of hormones from one tissue to another.

2. **Abscission**

Except the physical reasons such as wind or predator damage, the fruit as consequence of reduction in hormonal concentration in fruits particularly in the specialized layer of cells between fruit and stem or branch called abscission zone/layer. Abscission is associated with June/December drop can be reduced by hormonal application to the fruits. Auxins like NAA act as "stop drop chemical" in low concentration and chemical thin the higher concentration.

3. **Mobility of Photo-synthates to the Fruit**

The translocation of carbohydrates takes place from leaves to fruits against concentration gradient mainly due to some enzymatic activity otherwise fruit should not have been able to store large quantities of fats, carbohydrates *etc.* Hence proper photosynthesis is very important for fruit development. So, process including pollination, fertilization, seed setting, number of seeds, as well as number of leaves (for photosynthesis) influence fruit development.

In seedless fruits – it is assumed that some other parts of fruit may take up the function of seeds including hormonal action.

External Factors Affecting Fruit Growth

Despite overall importance of hormones, role of photosynthesis in the leaves, water and nutrients transported from soil through roots, light and plant health must not be forgotten. Of course, most of these factors depends on grower:

Water: If supply of water is short the fruit may go through its normal cycle of development but will not grow to its optimum size.

Nutrients:

Adequate nutrient supply: Healthy plant with plenty of healthy leaves – more photosynthesis- more carbohydrates production – improve flower and fruit quality.

Low level of nutrients may directly influence flower and fruit development.

Deficiency of boron: Cause cracking in litchi, pomegranate, lemon.

Deficiency of calcium: Bitter pit in apple.

Deficiency of potassium: Reduced colour and sweetness.

Deficiency of nitrogen: Small size fruits.

High nitrogen: Poor storage life.

Light: Although a fruit growing in abundant sun light yet problem is of exposing maximum number of leaves to direct sun light. So dwarfing root stock (to avoid internal shedding), training and pruning is practiced for more carbohydrate production by photosynthesis for better fruit growth.

Pest and diseases: Mark on fruits by insects/diseases, decrease number of leaves which reduce photosynthesis and carbohydrate production leads to fruit drop and decreased fruit quality and size.

Fruit Maturity

It is the time between final stage of growth and beginning of ripening.

Maturity Indices of Fruits

- **Maturity (Commercial or Horticultural maturity):** It refers to the stage at which the produce is optimally accepted by the consumers.
- **Maturity index:** The maturity index at harvest determines on the basis of skin colour, shape, size, aroma chlorophyll and firmness for quality and post- harvest shelf-life of the fresh fruits.
- Maturity indices helps to harvest the crop at right time.

There are measurable points which gives notion of harvesting to ensure proper quality of produce. Some commonly used maturity indices are:

- Attainment of proper shape and size.
- Change of peel colour to greenish yellow, purple red, orange, brown as per characteristics of variety.
- Change of pulp colour
- Decrease in firmness of fruits
- Decrease in starch content
- Decrease in acidity

- ☆ Increase in sugar

Judging Maturity

- ☆ There are various means of judging maturity but they vary according to the kind of fruits, local soil and climatic conditions but generally farmers follow visual means (appearance). However, this is not a perfect method, many a times it may mislead the farmers.
- ☆ There are certain limitations for maturity indices due to variation in nutrition of the crop, fruit shape, size, climate, seasonal factors, moisture, pruning method, use of hormones and other chemicals.

Estimation of Maturity

Different methods are deployed to judge maturity of fruits. These are as under:

i) Computation Method

- ☆ The calendar date is used to judge maturity for fruit crops grown in seasonal climate which are almost uniform from year to year.
- ☆ Day to bloom.
- ☆ Heat unit accumulation.
- ☆ Angle between fruit receptacle and peduncle when it reached about 90°.
- ☆ However, these factors may vary with the change in climatic conditions.

ii) Physical Method

This method based on discernible changes that takes place in fruit at the stage of harvesting. Shape (banana - angularity of fruit surface disappear and becomes round). In pineapple, the eyelets flattened in the centre and bulge on the side at ripening, size (banana, litchi), weight, surface texture (scale on the surface disappear in sapota and fruit turn smooth), specific gravity (mango - judged by putting fruit in water), colour of exocarp, flesh firmness (penetrometer), tapping sound (dull metallic sound of water melon on ripening), colour of leaves (leaf fall in bael), natural detachment (*tapka* in mango) *etc.*

iii) Change in Colour Method

Surface colour undergoes sharp changes with maturity. Largely green colour in fruits during early growth stage owes to the presence of chlorophyll changes to carotenoid/anthocyanin pigment during maturity. It may be changed from deep green to light green or may be development of yellow or red or purple pigment or blush colour *e.g.* harvesting of papaya at 2/3rd fruit is yellow stage gives quick ripening. The colouring pigments in fruits, Carotenoid (mango, citrus, banana, sapota, pineapple), carotenoid- caricaxanthin (papaya), carotenoid, lycopene - guava, carotenoid, cryptoxanthin (avocado).

iv) Physiological Method

It is based on ethylene evolution in climacteric fruits. This method is applicable in climacteric fruits only. Fruits are classified as:

Table 8.2: Concentration of Ethylene Required for Fruit Ripening

Ethylene Evolution Class	*Range at 20°C μ_1 C_2H_4 per kg per h*	*Examples*
Very high	More than 100	Passion fruit, sapota
High	10-100	Avocado, papaya
Moderate	1-10	Banana, fig, guava, mango
Low	0.1-1	Pineapple
Very low	Less than 0.1	Citrus, grape, pomegranate

v) Chemical Method

- ✰ Chemical characteristics is related to the palatability of the produce. Obviously, it is important characteristics to determine maturity.
- ✰ Conversion of starch to sugar is used to judge maturity in apple. It is based on reaction between starch and iodine which produce a blue or purple colour. It indicates the amount of starch remaining in the fruit.
- ✰ Carbohydrate and organic acids breakdown into sugars during ripening. Increase in TSS, sugars and aroma compounds indicate ripening.

Ripening

The growth and development in which transformation of fruits from an unfavourable state of firmness, texture, colour, flavor and aroma to a more favourable state of consumption.

Role of Hormones in Ripening

Ethylene (Ripening hormones): When ethylene production inhibited the ripening of fruits is delayed. An increase in ethylene in non-climacteric fruits like citrus, usually does enhance respiration rate.

Attached fruit: Release of ethylene is slow as leaves produce an ethylene inhibitor.

Detached fruits: Ethylene production at rapid rate.

Auxin: Treatment with an auxin (2,4-D) stimulates maturation and ripening through stimulation of ethylene production.

Gibberellins: GA delay ripening and reduce respiration. GA and ethylene have opposition effects.

Climacteric fruits: There is rapid rise in respiration rate during ripening. Ethylene lead to early ripening of fruits *e.g.* banana, mango, sapota, papaya, guava *etc.*

Non-Climacteric fruits: Respiration rate after harvest remain very slow. So, ripening takes place on the tree itself *e.g.* grapes, citrus, litchi.

Physiological Changes Accompanying Senescence of Horticultural Crops

Cellular

- ☆ Loss of chlorophyll, disassembly of chloroplast structure.
- ☆ Degradation of cell walls.
- ☆ Altered membrane composition, loss of fluidity.
- ☆ Loss of cellular compartmentation, release of vascular contents.

Composition

- ☆ Altered sugar content and switch to alternative substrates for respiration.
- ☆ Net loss of RNA.
- ☆ Increased protease activity, net loss of protein.
- ☆ Altered amino acid content.

Physical Changes Accompanying Senescence of Horticultural Crops

- ☆ Colour: Loss of green colour, synthesis of new pigments (carotenoids, flavanoids).
- ☆ Texture: Softening, wilting, drying.
- ☆ Loss of resistance to pathogens: Development of infections.

Sensory attributes: Loss of appearance (shape, size, colour), taste (sweet, salty, bitter, sour), odor/smell/aroma, texture/mouth feel.

Harvesting

Harvesting is a deliberate operation to separate the produce from source of its attachment. Shelf life, transportation, marketing, processing, value addition and ultimately consumers satisfaction is concerned depending upon harvest time/stage. Harvesting at immature stage and following improper method lead to damage of produce and producer both. Harvesting depends upon many factors such as nature and type of crop. Delicacy of crop, maturity criteria, time of harvesting mode of transportation, purpose of use, cost economy *etc.*

Types of Harvesting

1. **Manual harvesting**: Scissors, sickle.
2. **Mechanical harvesting**: Tree is shaken by mechanical vibration.
3. **Chemical harvesting**: Some chemicals are sprayed on the tree before

harvesting to loosen the attachment of fruits to the plant. MH@ 1000ppm delayed the ripening of hard green mangoes during storage.

Mango

- Slight colour development on the shoulders.
- When one or two ripe fruits fall from the tree naturally (*tapka* method).
- Counting of the days from fruit set to maturity.
- When the specific gravity of fruits ranges between 1.01 to 1.02.
- Flesh firmness.
- When the TSS reaches 11-15°Brix.

Banana

- The fruits are harvested at different maturity level based on distance of transport.
- For long distance transport fruit can be harvested at 75-80 per cent maturity and for short distance transport 90-100 per cent maturity.
- Days taken from shooting *i.e.* 3.0-3.5 months.
- Disappearance of angularity of the fingers.
- Brittleness of floral ends.
- Drying of leaves in some varieties.

Guava

- Guava fruits generally take about 17-20 weeks from fruit set to reach maturity.
- When the colour changes from dark green to light green.
- When the specific gravity is one (1.0).

Grape

- Grape is harvested when they reach a TSS of 16 to 24 per cent depending on variety.
- Bangalore Blue: 12-14 per cent.
- Anab-e-Shahi and Selection-7: 16-18 per cent.
- Thompson seedless: 20-22 per cent.
- Besides TSS, the following physical characteristics are also useful in judging maturity.
    - Texture of pulp (softness).
    - Peel colour (light yellow).
    - Easy separation of berries from bunch.

- ✫ Development of characteristic flavour and aroma.

Papaya

- ✫ For local market: When skin colour changes at the apical end of the fruit.
- ✫ For long distance transport: When the skin colour changes from green to yellow (to the extent of 6 per cent).
- ✫ When the latex of the fruit becomes almost watery.

Pineapple

- ✫ For local market: When 25 per cent of surface changes to yellow colour.
- ✫ For long distance: When all the eyes are still green and have no trace of yellow colour (75-80 per cent maturity).

In India

1. Pineapples are harvested when colour changes from green to greenish yellow.
2. The fruit develops a smooth surface around the eyes.
3. The flattening of eyes.
4. TSS: acid ratio of 21 to 27 and specific gravity of 0.98 to 1.02.

Jackfruit

- ✫ A dull, hollow sound is produced when the fruit is tapped by the finger.
- ✫ The last leaf of peduncle turns to yellow.
- ✫ Fruit spines become well developed and widely spaced.
- ✫ An aromatic odour develops.

Pomegranate

- ✫ The computation method for fruits harvest of pomegranate fruit is between 135-170 days after anthesis.
- ✫ The skin colour of fruit changes to dark yellow in summer and dark red in winter.
- ✫ The persistent calyx at the anterior end of the fruit curves inward and become hard and dry at maturity.
- ✫ Rind is very hard.

Sapota

- ✫ The peel shows a dull orange or potato colour with a yellowish tinge when scraped.
- ✫ The scurf content on the surface of the fruit will be minimum and easily fall off.
- ✫ The content of milky latex drops to almost zero.

Citrus

- ☆ Maturity indices differ among the citrus species/varieties.
- ☆ Mandarins: When the rind colour changes from green to orange.
- ☆ Sweet orange: When the rind colour turns to yellow.
- ☆ Limes: When the rind colour changes to light green to yellow.
- ☆ The International Standards Organization (ISO) has set in minimum juice content of citrus as follows:
  a. Washington navel oranges 30 per cent.
  b. Other orange varieties 35 per cent.
  c. Grapefruit 35 per cent.
  d. Mandarin orange 33 per cent.
  e. Lemons and limes 25 per cent.
  f. For processing total juice content of the fruit is important.

Fig

- ☆ When fruits become soft and wilt at neck.
- ☆ Fruits hanging down from their own weight.
- ☆ No milk exudation from the stem when the fruit is pulled off.

Custard Apple

- ☆ When the fruit turns to light green colour.
- ☆ Development of yellowish white colour between the carpels.
- ☆ Initiation of widening the gap between carpels or segments.

Ber

- ☆ Attainment of full size, soften of pulp and change in colour from greenish to golden yellow.
- ☆ Days to mature the fruits *e.g.* in Delhi cv. Gola took only 150 days where as Tikadi requires 173 days to mature.
- ☆ In cultivars like Gola, Kaithali and Umran, there was an increase in reducing and non-reducing sugars, total sugars and TSS.

Date Palm

Table 8.3: Fruit Development Stages of Date Palm

Sr. No.	*Stages*	*Characteristics*
1.	Hababouk stage	It is the initial stage of fruit development usually it takes about 4-5 weeks. Fruit weight is about one gram and fruits are covered by calyx.
2.	Kimri (Jimri) stage	The fruits are hard, green in colour and are not suitable for eating. Fruits are acidic, stats accumulating sugars and gain weight at fast rate.
3.	Doka stage (khalal)	Fruit becomes hard, yellow or pink or red, TSS 30 to 45 °B astringency present or absent depending on cultivar and edible stage.
4.	Dang stage (rutab)	Softness starting at tip of fruit, tannins and astringency disappears, lose weight and moisture content is about 35 - 40 °B and edible stage.
5.	Pind stage (tamur)	Fully ripe fruit lose weight, TSS 60 to 84 °B and edible stage.

Chapter 9

Dormancy

Suspension in growth during the development of the plant. Sometimes it causes formation of special structures like tuber, bulbs, seeds and buds *etc.* This suspension of growth may due to internal (physiological) and external (environmental) factors. Such a state is called dormancy.

In case of seeds, if these may germinate upon exposure to favourable conditions, the seeds are said to be quiescent.

It is a condition in which seeds are unable to germinate when suitable environmental conditions (water, temperature and aeration) are permissive for germination, the seeds are said to be in rest period or dormant.

This situation happens in plant to survive against adverse environmental conditions plants undergo a period temporary growth suspension/dormancy with suitable modifications. Lower plants avoid these unfavourable conditions through the production of endosperms, zygospores, auxospores, akinetes *etc.* The vascular plants produce dormant structure to overcome such unfavourable conditions.

Among different plants structures seed and bud are important which exhibit dormancy. Extreme variation in the temperature becomes difficult for survival/ growth of plants, especially cold conditions. In order to survive against such adverse conditions, the growth regions like apical buds, axillary buds and underground structures like rhizomes, tubers *etc.*, undergo a period of dormancy it is also called as suspended growth.

Classification

Lang *et al.* (1987) classified dormancy into following categories:

1. Paradormancy

The inhibition of growth due to physiological factors or biochemical signals originated outside the affected structure *i.e.* apical dominance, photoperiodic response, bud scale effects. Paradormany encompasses shallow dormancy, summer dormancy, temporary dormancy *etc.* Basipetal transported auxins have been observed by many workers as a primary signal that regulates paradormancy.

2. Endodormancy

The inhibition of growth regulated by physiological factors inside the affected structure *i.e.* chilling response, photoperiodic responses. Endodormancy encompasses after ripening, deep dormancy, true dormancy, deep rest dormancy, rest *etc.*

Physiological changes internal to the bud that prevent untimely growth during seasonal transitions, when environmental conditions often fluctuate between those permissive or inhibitory to growth lead to endodormancy.

It helps to protect the vegetative bud by suspending the growth of meristem until return to suitable environmental conditions.

Induction and breaking of endodormancy is regulated by light and temperature. Light plays an important role in dormancy in woody plants. In deciduous plants, terminal buds produce leaf primodium forming scales instead of leaf bud due to short-day conditions.

3. Ecodormancy

The inhibition of growth by unfavourable environmental conditions *i.e.* extreme temperature, water stress, nutrient deficiency.

These environmentally unfavourable growing conditions are generally required for breaking of endodormancy but at the same time it may imposing dormancy.

Endodormancy encompasses after rest, post dormancy, environmental dormancy, relative dormancy, imposed dormancy *etc.*

Bud Dormancy

It is temporary suspension growth or physiological activities of bud during the adverse environmental conditions (heat, cold, drought) It can be resumed later on.

Or

During the developmental cycle certain parts of plant like buds, bulb, tubers, seeds *etc.*, go through rest period or temporary suspension of growth at some phase or the other such state is called dormancy.

Factors Affecting Bud Dormancy

1. External/environmental: Temperature, photoperiod, restored O_2 supply *etc.*

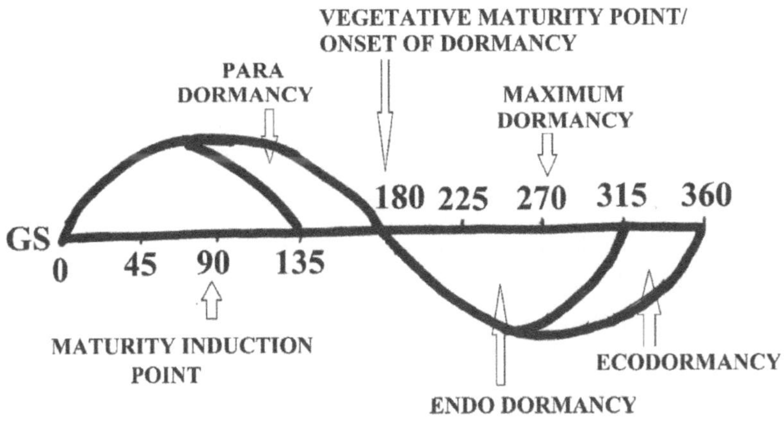

Fig 9.1: Representation of Dormancy

2. Internal (Hormonal): ABA, gibberellins, auxins, cytokinins and ethylene.
3. Nutritional: Nutrients diversion, Ca^{2+}
4. Genetic factors

Dormancy Controlling Factors

Organization level	Term for event
1. Endorormancy	**Type of dormancy**
a) Chilling	Caryogenic endodormancy
b) Photoperiod	Photoperiodic endodormancy
c) Endogenious	Rhythmic endodormancy
2. Paradormancy	
a) Apical meristem	Apical paradormancy
b) Seed coat	Testa paradormancy
c) Periderm	Peridermal paradormancy
d) Chilling	Cyrogenic paradormancy
e) Photoperiod	Photoperiod paradormancy
3. Ecodormancy	
a) Temperature	Thermal ecodormancy
b) Water	Hydration ecodormancy
c) Nutrient	Nutritional ecodormancy
d) Levels of CO_2, O_2	Atmospheric ecodormancy

(Source: Lang et al., 1987)

Categories of Seed Dormancy

I. Primary Dormancy

A. Exogenous Dormancy

The factors outside the embryo or affected part.

1. **Physical**: Seed coats are impermeable to water.
2. **Mechanical**: Seed coats are too hard to allow the embryo to expand during germination *e.g.* olive and ber.
3. **Chemical**: It may be due to some chemical inhibitors present in the outer coverings of seeds or fruits. *e.g.* capsules, apples, grapes, citrus, desert plants.

Overcome: Scarification – removal of seed covering mechanically or chemically, warm or cold stratification, removal of seed covering (fruits) or leaching seeds.

B. Endogenous Dormancy

The factors responsible for dormancy are present within embryo.

1. **Morphological**: Means at ripening of fruit seed is not developed fully, it requires some additional time after separation from plant for proper growth of embryo.
   a. **Rudimentary**: May be due to presence of inhibitors at pre-embryo stage.
   b. **Linear**: Takes up about a 1/2 of seed cavity. May be due to presence of inhibitors.
   c. **Undifferentiated**: Orchids
2. **Physiological**: When the seed is requiring some storage period for proper germination, it varies from 1 to 6 months.
   a. **Non-deep**: Short-termed dormancy and resumed after storage.
   b. **Intermediate**: The dormancy may be due to the seed coat or tissue surrounding embryo or within *e.g.* conifers.
   c. **Deep**: Controls are within the embryo itself *e.g.* temperate trees.

Overcome: A period of storage, exposure to red light for moderate period, alternate warm and cool temperatures stratification; also, hormones such as GA_3. Exposure to cool temperatures (59°F).

C. Combinational (two kinds of dormancy)

1. Morphophysiological

a. **Epicotyl**: Separate after-ripening requirements for epicotyl, radicle, hypocotyl.

b. **Epicotyl and radicle (double dormancy)**: Require chilling period for embryo, followed by a warm period for rooting, then a second cold period for shoot growth.

2. **Exo-endodormancy**

Due to combination of both endogenous and exogenous dormancy.

Overcome: Sequential combinations of dormancy releasing treatments cycle of warm and cold stratification. Example: Scarification followed by cold stratification.

II. **Secondary Dormancy**

Imposed dormancy mechanism during unfavorable conditions.

1. **Thermodormancy**: High temperature induces dormancy.
2. **Conditional dormancy**: Changes in ability to germinate, related to time of the years.

Advantages of Dormancy

1. Seed can be stored for longer period in seed bank.
2. Assured plant population can be maintained easily.
3. Seed can be easily dispersed by passing through a bird's and animal's digestive tract.
4. For temperate plants it is helpful for survival of plants against adverse environmental conditions.
5. Under tropical conditions survival of seed under water stress conditions due to impermeable seed coat.
6. Seed dormancy is important for storage of grains for a longer period. It also become easy to sow the seeds in the ensuing season.

Disadvantages

1. Need long time to overcome dormancy so variation in germination time.
2. Dormancy contributes to maintain weed seed bank in soil for longer time.
3. At the time of sowing it become very difficult to maintain appropriate population and or fix the seed rate.

How to Overcome Dormancy

Seed may remain dormant due to different kind of dormancy (hard seed coat, physiological dormancy) which can be overcome by scarification and stratification treatments.

1. **Scarification Treatment**

Removal of hard seed coat/covering by physical or chemical to enable them to germinate.

i. Acid ii. Mechanical iii. Physical treatment – hot water treatment

a) Acid scarification

Dip the seed in concentric sulphuric acid for 2-5 minutes, duration of dip varies from crop to crop or type of seed.

e.g. dip ber seed in concentric sulphuric acid for 6 minutes to remove the dormancy it enhances germination.

Tree crops 1-3 hours.

Rose partially dip in acid followed by warm water stratification.

b) Mechanical scarification

Process of removing hard seed cover by rubbing on sand or puncture seed coat with hammer to make it permeable to exchange of water and gases for effective germination.

e.g. removal of ber seed coat with hammer.

Rub bitter guard seed in 1:2 ratio of sand and seed, followed by rub on hard surface for 5-10 minutes.

2. Hot Water Treatments

Impermeable seed coats can be softened by hot water scarification. Seeds are placed in hot water at 77 to 100°C and are immediately removed from hot water, followed by soaking gradually in cool water for 12-24 hours. However, in Bengal gram and groundnut hot water dip for more than 1 minute is injurious to the seeds.

3. Stratification Treatment

It is pretreatment of the seed to stimulate the natural conditions for better germination *e.g.* seed dormancy is due to embryo factor.

a) Cold stratification

Low temperature treatment to the seeds (0-5 °C) for germination for a period of 2-3 days to several months. It varies with crop and type of dormancy in the seed *e.g.* cherry and oil palm seeds.

b) Warm stratification

Some seed required to expose in high temperature for breaking dormancy. The moisture content during temperature treatment should not be more than 15 per cent.

e.g. paddy seed exposed at 40-50°C for few days to remove dormancy.

Oil palm seed exposed at 40-50°C for 2 months to remove dormancy.

4. Leaching of Metabolites (Inhibitors) Treatment

It is used to leach down the dormancy promoting chemicals or inhibitors by dipping the seeds in fresh water for 3 days but water should be replaced once in 12 hours to avoid fungal infection or can be dipped in running water.

Coriander to leach coumarin.

Sunflower to leach hydrocyanic acid.

5. Chemical Treatment

- ✰ Soak freshly extracted dormant seeds in potassium nitrate (0.2 per cent) for improving seed germination.
- ✰ Peaches and grapes seeds treated with Thiourea @ 5000ppm improved germination.
- ✰ The oxygen deprivation worked even more than a cold treatment for dormancy release in apple and peach.
- ✰ Treat/dip ber seed in concentrated sulphuric acid for 6 minutes to remove/burn the hard seed coat.

6. Growth Regulators Treatment

In case of inhibitors (ABA) induced endogenous dormancy in the seed apply growth regulators at low concentration.

- ✰ GA (100-500ppm) can be used to break dormancy in temperate fruits.
- ✰ Cytokinin and kinetin @ 100-1000 ppm used to break dormancy.
- ✰ Ethrel (5000ppm) can be used for breaking the dormancy of guava and strawberry.
- ✰ Benzyladenine (BA) @10-20 ppm is found effective in breaking seed dormancy in apple and peach.

7. Temperature Treatments

a) Low temperature treatments

Chilling requirement for plants grown under temperate conditions.

e.g. seed dormancy of apple can be released by storing seeds at 5°C low temperature treatment.

b) High temperature treatment

Normally high temperature treatments are exhibited by early flowering "winter" annuals.

e.g. blue bell (*Hyacinthoides non scripta*) seeds require high temperature summer heat for germination.

c) Alternate temperature treatments

The plants grown in cool or temperate zone require alternate temperature for dormancy breaking *e.g.* bull rush (Typha).

8. **Light and Phytochrome Treatment**

Red light (670 nm)

Inactive form of Pr ⇄ Pfr Active form of Phytochrome (non dormant)

Phytochrome (Dormant) (Yellowish green in colour)

(Blue green in colour) Far red light (730 nm)

Red light promotes germination and far red light affects germination.

9. **Promoters - Inhibitors Treatment**

For proper germination promoters and inhibitors present in the seed should be in a balanced manner.

- ✰ GA helps in translocation of reserve food materials to active site of meristematic activity as well as cell division.
- ✰ Cytokinin being endogenous hormone controls germination through DNA to RNA transcription system.
- ✰ Abscisic acid affecting RNA synthesis to prevent seed germination.

Infra-red radiation treatment: Infra-red is also helpful to release dormancy.

Pressure Treatment

Dormancy of seeds can also be broken by pressure treatment in autoclave.

Chilling

Chilling Requirement

The low temperature requirement of a plant to resume its normal growth during spring season is commonly referred to as its "chilling requirement." Low temperature requirement, ranges from 3 to 7°C for a particular period to break the dormancy. It helps to increase promoters' level (GA, cytokinin *etc.*) and decrease inhibitors level (ABA).

Certain fruit species require low temperature exposure (chilling) for normal and abundant bud break during spring. In certain years, insufficient chilling has been recognized as an economic loss in the production of several temperate zone fruits, particularly those temperate fruits which are grown in sub temperate climates. There are so many symptoms of insufficient chilling but most common are delay in flowering and vegetative bud break, bud break spread over an extended period of time, or in several cases a total lack of bud break. In insufficiently chilled tree, it is difficult to find large fruit and flowers on the same shoot as well as sparse bud break levels.

What Happens during Winter Chilling

During low temperature or winter temperate the plants goes into dormant period, it is also called as rest period. Leaf fall and cease visible growth occur when air temperature fall below 50 °F and plants enter the rest or dormant period. During this phase of growth, inhibitor increase and growth promoter decrease. As the chilling requirement of plant satisfy the growth inhibitor decrease or leach down while growth promoter increase. Increase in level of growth promoter allow/ resume normal growth of plant and start flowering after fulfilling the chilling requirement.

Measuring Winter Chilling

Chilling hours: The term was coined by Weinberger (1950). Amount of chilling required by fruit plants to break dormancy.

Chill unit: Hours of temperature between 0-10°C required to break endodormancy (regulated by physiological factors inside the affected structure).

One chill unit: Exposure of plant for one hour at optimum temperature.

The temperature between 32 to 55 °F provide the chilling effect on the plant, however optimum temperature to provide chill unit is 45 °F.

Temperature below 32°F contribute nothing or little to receive chill unit and the temperature above 70°F for more than 4 hours negates the chill unit already received by the plants during last 24 to 36 hours.

Number of models have been developed by different researchers to measure the accumulation of chill unit for satisfying the rest period. Each new modal has the improvement over the existing model by monitoring daily temperature of 45 °F or lower from October one each year.

Among different models Alabama, the Modified 45 provided best to predict the satisfaction of the rest period by cold temperature. This model counts the chill unit when the temperature is below 32 °F and does not have negative effect of higher temperature over the accumulated unit. This method actually measures the rest and start date of fall in temperature at or below 45 °F instead of October 1. This method is superior over the Utah, the Florida and old 45 method for measuring chill unit under Alabama conditions.

Table 8.1: Chilling Unit Model – by Richadson *et al.*, 1974

Temperature	*Chill Unit*	*Temperature*	*Chill Unit*
<1.4 °C	0.0	1.5-2.4 °C	0.5
2.5-9.1 °C	1.0	9.2-12.4 °C	0.5
12.5-15.9 °C	0.0	16-18 °C	-0.5
18.1-21 °C	-1.0	21.1-23 °C	-2.0

Table 8.2: Chilling Requirements for Fruit Species

Types Fruit Type	*Chilling Hours Requirement (Temperature < 7°C)*
Apple	200 to 2000
Apricot	250-900
Domestic plum	700-1700
Japanese plum	500-1500
Peach	200 -1100
Quince	50-400
Sour cherry	600-1500
Sweet cherry	500-1300
Pear	850-2600
Almond	100-700

Low Chilling Varieties

Apple: Anna, Dorsett Golden, Ein Shemer, Tropical Beauty, Gordon, Winter Banana.

Pear: Hengshan, Hood, Flordahome, Pineapple, Tenn, Seleta, Baldwin, Orient, Kieffer.

Peach: Bonita, Desert Gold, Early Amber, Fairtime, Gold Dust, Tropic Sweet, Indian Blood, August Pride.

Apricot: Royal, Perfection, Newcastle, Katy, Garden Annie, Gold Kist.

Plum: Burgundy, Delight, Howard Miracle, Mariposa, Satruma, Santa Rosa.

Importance of Chilling/Causes of Insufficient Chilling

1. **Delayed foliation**: Insufficient chilling resulted in delayed foliation. Trees may have small tuft of leaves 10-20 inch below the tip. Lower buds will break, foliation delayed, reduced fruit set and tree will become weak. Further, suckers will develop from the lower portion of the tree which create management problem.
2. **Reduced fruit set and buttoning**: When chilling requirement is not fulfilled properly or insufficient chilling units are accumulated, flowering is delayed and extended. It also results in poor fruit set due to abnormalities in pollen and pistil development. In many peach cultivar flowers drop before or around shuck split, but in others such as 'Jersey Queen' and 'Harvester', buttons form. Buttons results from flowers which apparently have set but never develop into full size fruit. The fruit remains small and misshapen as they ripen. When we cut these fruits there will be no seed or dead seed. Because buttoning is not developed in early season. These fruits will act as food source and over wintering site for insects.

3. **Reduced fruit quality**: Insufficient chilling requirement in fruits is not much studied, however, it is major problem. Insufficient chilling affects the size and quality of fruits. In some cultivars caused enlarged tip of the fruits and reduce firmness of the fruit. Ground colour of the fruit remain green and it loose firmness before the ground colour change from green to yellow. These varies with the cultivar and the amount of insufficient chilling.

Length of Chilling Requirement in Fruit Species

There are some observable symptoms in fruit trees that have chilling hours incompatible with the local climate. Fruits with low chilling requirement affected with frost during flowering due to early flowering and late spring, it also damages the crop. In these situations, chilling requirement exceeds and result in uneven bloom and foliation. If high chilling requirement varieties are planted under insufficient chilling conditions these are unable to flower/bloom or foliation. Chilling requirement is estimated on the basis of average from extremes and each chilling requirement has some latitude.

In general, for elevations above 6000 ft, varieties with chilling requirements above 1000 hours, for elevations between 4000-6000 ft, varieties with chilling requirements between 750 and 1000 hours and the elevations between 2500-4000 ft varieties with chilling requirement between 500-750 hours should be used. There is some flexibility depending on orchard site/micro climate.

Rootstocks

The chilling requirement of a root stock can have an effect on chilling requirement of the variety. Most of the clonal apple root stocks used commercially were selected and bred in much colder climates, so how they work in warm area is not been fully tested.

Modification of Chilling Requirement

A. Environmental Factors

1. Pre-chilling Growth Promotion

- ✰ Extended growth period of shoot may delay bud break in the following spring. Shortened growth period of shoots caused by environmental factors might reduce the chilling requirement.
- ✰ Root stock may also influence the chilling requirement of scion in some species *e.g.* pear, peaches, almond and apple. However, nature of this influence has not been established.
- ✰ The leaves retarded dormancy release because they found that longer leaves remained on the tree in autumn, the dormancy was longer and deeper.

☆ In tropics and sub tropics, artificial defoliation is an efficient method to induce floral bud break in deciduous fruit trees.

2. Intermittent Warm Period

High temperature during winter is well known to prolong dormancy. But when temperature become exceptionally high *i.e.* 35°C and above, they may lose their inhibiting effect and instead of delaying may promote budbreak. Both treatments (warm water treatment at 45°C for 3.5 h and chilling at 3°C for 8-12 weeks) causes bud break in pear but bud activity was much more pronounced and the sprouting more vigorous in the cold treated plants.

In apple trees, heating can replace the chilling requirement only if buds are in a light state of dormancy whereas bud in deep dormancy did not respond.

3. Radiation

Generally chilled plants started into growth promptly, whether the chilling was done in full light, in partial light or in complete darkness. Nectarine trees covered with 55 per cent shed saran cloth terminated their dormancy 10 days earlier than un-shaded plants. The low light intensity always significantly promoted floral bud break in peach.

4. Leaching

The leaching of inhibiting substances by rain is an important factor for reducing chilling requirement of buds. The effect of rain to be caused by reduced bud temperature rather than leaching.

B. Cultural Practices

1. Reducing chilling requirement

Since deciduous fruit trees are grown in a subtropical and even tropical regions, many attempts have been made to overcome the problem of insufficient chilling. Basically, two approaches are possible.

a. **Preventing plant from entering dormancy** – This is feasible only in regions without distinct seasons, where plant do not need a dormant period in order to survive adverse environmental conditions. The dormancy can be prevented by defoliating trees before entering into endodormancy.

   The leaves may be removed by mechanical or chemical means by extending drought.

b. **Hastening dormancy release** – After plants have entered dormancy, it can be released by several methods:

   i. Increasing chilling through evaporative cooling.

   ii. Heat treatment at temperature 35 °C to 50°C.

iii. Late autumn application of N and irrigations.

iv. Treatment with several chemicals.

Table 9.3: Compounds for Breaking Dormancy in Fruit Trees

Sr. No.	*Compound*	*Concentration Used*	*Remarks*
1	Dinitro-ortho-cresol (DNOC)	1.5 per cent in mineral oil as dormant spray	-
2	KNO_3	2-5 per cent	Promote more flower bud opening
3	Thiourea	2 per cent not closer than 2 weeks to bud break	Promote more leaf bud opening
4	Benzyl adenine	500ppm	Dose not translocate, growth limited
5	Gibberellins	50-200 ppm	Most effective on sour cherry and peach

2. Increase of Chilling Requirement

Delay the blossom period to avoid damage from late spring frosts have been made by delaying bud development after the termination of true dormancy. An example of this is evaporative cooling by sprinkling, which causes some kind of imposed dormancy by delaying heat unit accumulations. There is possibility of retarding bud break by autumn application of GA which prolong dormancy of bud.

Table 9.4: Chilling Unit Models Given by Different Scientists

Chill Unit Value	*Corresponding Temperature (°C)*		
	A	*B*	*C*
0	<1.4 °C	-1.0 °C	-1.1 °C
0.5	1.5-2.4 °C	1.8 °C	1.6 °C
1.0	2.5-9.1 °C	8.0 °C	7.2 °C
0.5	9.2-12.4 °C	14.7 °C	13.0 °C
0	12.5-15.9 °C	17.0 °C	16.5 °C
-0.5	16-18 °C	19.5 °C	19.0 °C
-1.0	>18 °C	21.5 °C	20.7 °C
-1.5	-	-	22.1 °C
-2.0	-	-	23.3 °C

A: Richardron *et al.* (1974) in high chilling peaches; B: Gilreath and Buchanan (1981), in low chilling nectarine; C: Shaltout and Unrath (1983), in high chilling apple.

The disadvantage of all these models is that these are based on the influence of temperature which is imperfectly understood and that they ignore the action of other environmental factors known to modify the chilling requirement. Accordingly, they cannot predict dormancy release accurately under all conditions.

How to grow temperate fruits in low latitude?

- Adopt low chilling varieties.
- Adopt low chilling rootstocks.
- Spray suitable chemicals to reduce chilling requirement.
- Special pruning.
- Sprinkling over tree canopy.

Endogenous Control of Dormancy Release

1. Nutrients

Initial research indicated that insufficient supply of sugars being the cause of dormancy and release from dormancy by chilling being brought about by promotion of processes converting starch into sugars. But more recently it has been reported by several scientists that increase in sugar in the result rather than cause of dormancy release.

From the analysis of vegetative bud of pear, it has been reported increase in N, soluble P, DNA and RNA were not strongly related to dormancy release.

2. Oxygen

The rapidly of dormancy release was promotional to the degree of reduction in oxygen and to the duration of exposure to low oxygen in peach.

The dormancy of apple embryos could be broken by anaerobiosis without chilling. The oxygen deprivation worked even more than a cold treatment.

Several reports have been suggested that a reduced oxygen supply in seeds did not promote but rather prevented dormancy release.

3. Hormones

i) **Gibberellins:** The dormant unchilled seed have minute amount of GA, this increase during chilling process. GA has been thought to closely related to dormancy release.

GA treatment increase the dormancy breaking effect of DNOC fore and after deep dormancy. In seed GA promotes the biosynthesis of several hydrolytic enzymes (amylase, ribo-nucleases) that are required for germination process. GA could substitute for chilling. GA promote terminal bud relative to laterals which increased apical dominance. This does not occur if bud break is caused by sufficient chilling in peach and therefore it is unlikely that GA really can substitute for chilling during deep dormancy.

ii) **Cytokinin**: Cytokinin are known for the ability to induce cell division in certain plant tissue culture. Synthetic cytokinins are benzyl adenine (BA) and kinetin. In growing shoots, they are known to counteract the inhibition of lateral bud growth resulting from apical dominance. The involvement of cytokinin in dormancy release apparently starts only after some other prior changes are induced. Cytokinins were not or were partially effective without some prior cold treatment and cannot fully substitute for chilling.

iii) **Auxin:** Promote growth at low concentration, however, inhibits at high concentrations like 2,4-D. The concentration needed for inhibition to be lowest in roots, higher in buds and highest in stem.

Dormancy Models

While specific predictive bio-chemical marker events have yet to be characterized for bio-physical changes or gene expression during dormancy, models which predict bud break have been put forth.

1. Chilling Unit Model

Chilling temperature is required for the spring growth of dormant buds. Weinberger (1950) coined the term chilling hours for winter air temperatures less than 7°C and correlated with chilling hour (CH) to the amount of chilling required by fruit trees to break dormancy. Utah Model was proposed by Richardson which assigns chilling effectiveness value to various winter temperatures. One chilling unit (CU) is defined as one hour at optimum chilling temperature. The Utah model has been criticized for inaccurate predictions of plant response especially in areas with mild winter.

2. Degree Growth Stage Model

The degree growth stage model (GS Model) attempts to consider the changing physiology of plant during dormant period.

3. Two Step Cycled Temperature Model

A two-step chilling model was suggested to accommodate the cycling of chilling temperature with more moderate temperatures. The first step involves a precursor for the dormancy breaking factor and second step is needed irreversibly to process the transition from an unsuitable precursor to a stable dormancy breaking factor. The transition occurs only after a critical level of the precursor has accumulated.

Heat Unit

Heat Units: Quantitative term used to describe the accumulation of heat required for fruit to reach maturity and expressed in terms of temperature time values also called growing degree days (GDD).

Effective heat summation for a given location largely determines the length of time from bloom to ripening for a given fruit or variety.

Optimum temperature for photosynthesis is 70-80°F.

Traditional method of calculation using maximum, minimum temperature.

Degree Days calculation: (Max Temp + Min Temp)/2 - base temp = degree days example: (70+50)/2 - 50 = 10 degree days.

Used to predict bud break, harvest date, bloom date, vegetative growth, insect and disease growth (for IPM). It can vary with the stage of development.

Growing degree day or Heat unit: Certain amount of exposure to warm temperature more than 4.5 degree (Ullah Model of Richardson).

Growing Degree Days Requirements

Most of the fruit plant requires growing degree days (GDD) or heat unit for initiation of vegetative bud, floral bud or development of fruit. The requirement of the heat unit varies from crop to crop.

There are so many ways to calculate degree days but the simplest way is to calculate the average of maximum and minimum temperature and deduct the base temperature from average temperature.

The formula is then: GDD = (Tmax + Tmin)/2 - Tbase

As an example, if the daily maximum temperature is 75 °F and the daily minimum is 55 °F, then the mean temperature will be 65 °F. If the base temperature is 50 °F, then: GDD = 65-50 = 15 or 15 heat units.

If the daily minimum temperature is less than 50 °F, 50 °F is used. If the daily maximum is greater than 86 °F, 86 °F is used. So, let's consider a day where the high temperature is 60 °F and the low is 38 °F, average temperature would be calculated as 60+50/2 or 55 and GDD = 55-50 = 5. Similarly, if the high were 95 °F and the low 70 °F, average temperature would be calculated as 86+70/2 or 78 and GDD = 78-50 = 28 heat units or GDD.

Growing degree days is the heat accumulated as a sum of air temperature between 40 °F to 77 °F (or lower). If the air temperature remains below the 40 °F, the heat unit accumulation will be zero, and if the air temperature rises above 77 °F, it will be considered 77 °F to calculate the heat unit because no additional benefit was derived from the higher temperature (as per research model) *e.g.* if the air temperature is 39 °F the heat accumulation is zero; and If air temperature is 65 °F than heat accumulation will be 65-40=25 units; if air temperature rises to 80 °F than the heat accumulation will be 77-40 = 37 units. Warm day during summer can accumulate about 650 to 700 heat unit in 24 hours. Growing degree days or heat unit can be used to estimate the development stage of the crop.

$GDD = (T_{max} + T_{min})/2 - Tb$

T_{max}: Maximum ambient temperature.

T_{min}: Minimum ambient temperature.

Tb: Base temperature (generally 5 °C).

Early grape requires: 1600 heat unit to mature.

Late grape requires: 3500 heat unit to mature.

Based on growing degree values, the aonla cultivars can be classified in to two groups:

a) Early - Having GDD value between 5000 and 5200 degree days are Krishna, NA6 and NA7.

b) Late - Having GDD value more than 5200 degree days are Kanchan, Chakaiya and Anand-1.

Chapter 10

Fruit Drop

Shedding of fruits before harvesting or premature shedding/falling of fruits. It may be due to internal or external factors.

Internal Factors

a. Genetic factor
b. Hormonal imbalance
c. Morphological

External Factors

1. Biotic Factors

a. Insect pest and diseases
b. Damage by birds and animals *etc.*

2. Abiotic Factors

Climatic conditions (Wind speed, temperature, humidity, rainfall, hail storm *etc.*).

Stages of Fruit Drop

Stages are classified on the basis of maturity and size of fruit.

Post Bloom Stage

Fruit drop just after fruit set. About 80-90 per cent fruits drop at this stage. It may be due following reasons:

- ☆ Poor pollination.
- ☆ Shortage of nutrients.
- ☆ Adverse environmental conditions.
- ☆ Insect pest and diseases (fungal infection of *Colletotrichum acutatum*).
- ☆ Over production in the previous season.
- ☆ Non-availability of pollinator (bees) due to adverse weather conditions.
- ☆ Natural: Sometimes natural fruit drop at this stage is beneficial to reduce the excess fruit load.

June Drop

The fruits drop about one month after fruit set (in the month of June, so called June drop). Fruit drop at this stage is about 10 per cent.

Reasons

- ☆ It may be due to excess fruit setting.
- ☆ This drop may be due to competition among fruits for nutrition and carbohydrates.
- ☆ June drop in citrus may be due to water deficit, warm temperature.

Pre-harvest and Premature Fruit Drop

This is the last wave of fruit drop. This stage of fruits drop occurs just before the harvesting or attainment of maturity or horticulture maturity.

It occurs in mature fruits due to abscission of calyx abscission zone (button) or fruit drop without peduncle.

Reasons

- ☆ May be due to insect pest and diseases (fruit fly and fruit rot).
- ☆ Due to low temperature or foggy weather.
- ☆ Sometimes due to increase in ethylene content in the fruit at ripening stage which leads to production or formation of Abscission in the stem.

Causes of Fruit Drop

Pollination

i) Non pollination

ii) Over pollination

Non-pollination

Insufficient pollination may be due to non-availability or insufficient availability of pollinator (bees).

Remedies

- ☆ Avoid use of pesticides during flowering.
- ☆ Grow crops which attract insects specially pollinator.
- ☆ Avoids excessive irrigation during flowering.

Over Pollination

It may be due to over pollination or occurrence of more pollen grains on stigma (10-18 pollen grains).

This excessive pollen grains cause drying of stigma (about 93 per cent). Thereafter it causes drop of receptive female flower.

Non-Fertilization

(a) **Gametic sterility**: It is due polyploids levels, the uneven multiple of basic number of chromosome number fail to germinate pollen grains if it germinates than burst after reaching to ovary and production of imbalance embryo.

(b) **Incompatibility:** It is due to occurrence of physiological reactions between pollen tube, style and ovary tissues. The fruit drop just after fruit set in Blackcurrants is mainly due to auto-incompatibility of the varieties.

(c) **Failure of double fertilization**: When the sperm cell is not fused with poller nuclei prevent the endosperm formation. Preharvest fruit drop in Napoleon variety of sweet cherry due to degeneration of endosperm.

(d) **Abortion of embryo**: Embryo abortion may arise due to genetical and unfavourable nutritional conditions resulted in shedding of fruits. Higher fruit drop in late maturing sweet cherries is due to abortion of embryo. However, embryo abortion occurs often during the second phase of pericarp growth in super early cherries, but it occurs due to the competition for resources or food material between the seed and the pericarp.

Seed Content of Fruits

Synthesis of auxin in the endosperm of seed inhibit the formation of abscission. Auxin absorbs organic substances as well as influences the cytokinins distribution in the fruit, it also acts as sink of metabolites. So, the seed content in the fruit is most important precondition for the fruit to maintain it on the tree. Less seeds developing varieties/species are prone to fruit drop and susceptible to adverse environmental conditions such as water stress, poor nutrition *etc.* In case of apple when fruit set was abundant the fruits containing less than 3 seeds are shed first.

In case of dropped fruits the size of embryo stopped growing, it remains about one third of the normal size, however, in normal fruits it grows continuously.

Competition between the Organs of Plants

(a) **Competition among vegetative and reproductive organs:** The relation between vegetative growth and reproductive growth or fruit set is continuously changing during the growing season. If transport of photosynthates restrict the fruits will drop. The retention of fruits on the tree depends on ratio of leaves to fruits. In case of apple 1–4 leaves per fruit set are needed at the time of petals shed, 10–15 leaves per fruit are required around the June drop and 40 leaves per fruits are required at the end of fruit development. Leaf area is one of the important factors for fruit set, if inflorescences are more and leaf area is less than no fruit set or fruit will drop. More fruit drop is observed in the trees having strong shoot growth than the weak growing trees, it may be due to surplus fruit primordia on the fruiting structures. The other interesting observation is that the old leaves may stimulate fruit abscission, however, young leaves delay abscission of fruits. The old leaves translocate ABA from the leaves to fruits whereas young leaves delay the abscission of fruits.

(b) **Competition between the generative organs**: Plant having large number of flowers or fruit primordia causing a vigorous fruit drop because distribution of food is not optimum and resulted in low rate of fruit set. The flower or fruit set, which started growing earlier, becomes dominant in relation to other flower or fruits growing/developing later, this type of dominance is called primogenius. In case of apple, the apical flower start dominating due to earlier growth or starts growing first, however, in gooseberry inflorescence the flower at basal position develop first and start dominating others. The dominance is due to earliness rather than their position. The most developed flowers have the best chance to grow and are less exposed to be dropped.

Environmental and Physiological Fruit Drop

The physiological process of fruit drop is called abscission. Physiological drop occurs due to exposure of plant to environmental or physiological stress. The main cause of environmental and physiological fruit drop is:

Water stress: Due to competition, restricted amounts of water is one of the causes of fruit drop. It can be prevented by application of irrigation after fruit setting. Development of microclimate due to above crown spray of water as it increases the air humidity within the canopy, which is helpful in lowering the temperature of leaves and fruits effectively by evaporation thus reduces the chances of fruit drop. However, excessive watering in fruits like pomegranate, litchi, lemons, cherries *etc.*, may cause heavy fruit drop and develop cracks.

High and low temperature: Temperature has an important role on the vitality of embryo sac if the temperature is higher or lower than the normal it affects the longevity and vitality of embryo sac. Spring frost is threat for newly developed flowers in inflorescence of temperate crops because of flowers frost susceptibility during development. The flowers which open first are more susceptible to frost *e.g.* apical flowers in apple and basal flower in pear inflorescence opens first.

Nutrition: Balanced and timely application fertilizers is essential for normal growth and flowering of crop.

Nitrogen is one of the essential nutrients, it requires for normal growth and fruit set, it also reduce the June drops. However, excessive nitrogen causes fruit drop in blueberry as well as accumulation of poisonous nitrite also leads to fruit drop.

Boron plays important role in fruit set. Tree receiving high nitrogen also require high boron. However, more amount of K^+, Ca^{2+} and carbohydrate is needed for retention of fruits on the tree. Imbalance and improper nutrition also cause fruit drop due to poor health of tree.

Hormonal imbalance: Hormonal imbalance due to stress conditions, growth stage of the plant and other physiological factors also cause the fruit drop. Synthesis of ABA and ethylene during stress conditions cause fruit drop. Spray of 20 ppm 2,4-D during last week of April and 1st week of May is recommended to control fruit drop in Langra and Dushehari cultivar of mango.

Pathological Fruit Drop

Pathological fruit drop is a major bottleneck causing reduction in yield and quality of harvested produce. In this fruit drop, premature fruits fall down and give a painful sight to the growers. Such fruits become unfit for marketing. This drop is of much has economic importance and if left uncontrolled it reduce yield and returns.

Symptoms Besides causing fruit drop, pathogens are also responsible for die-back of twigs.

Attack of fruit borer and fruit fly also cause fruit drop.

Mechanical Causes

High velocity wind, hailstorm and inappropriate intercultural operation at the time of fruiting may cause fruit drop.

Cultural Practices

Cultural operations such as deep ploughing during fruit development stage and excessive irrigation (resulting bad aeration in soils) leads to drop of fruits.

Varietal Factors

Selection of the variety is one of the most important factors, varieties differ among themselves in the extent of fruit drop.

Chapter 11

Unfruitfulness

Problem of Unfruitfulness

Fruit trees fail to flowers regularly or equally and sometimes fail to flower or set the fruit under similar conditions where other trees fruiting well.

- **Fruit setting:** Start growth of ovary or its parts or developing after fertilization.
- **Fruitfulness:** It is a state of the plant to bear flowers, setting fruits and their maturity.
- **Infertility**: It is the inability of plants to produce fruit or viable seed. All fruitful plants are not fertile (seedless fruits) but all fertile plants are fruitful.
- **Self-fruitfulness**: Its ability of the plant to produce fruit or seed, these are self-pollinated plants.
- **Self-fertility:** It is capacity of a plant to produce viable seed or fruit after self-pollination.

Causes of Unfruitlessness

Unfruitful is a major problem in many orchards so the fact or factors responsible for unfruitful must be understood. The factors responsible for unfruitful are broadly divided into two groups:

- Internal factors
- External factors

1. Internal Factors Associated with Un-fruitlessness

There are so many internal factors responsible for unfruitfulness. Sometimes growers face difficult situation of poor cropping in his orchard. There may be abundant blossom but fail to set fruit. This is called unfruitfulness which may be due to following internal factors: i) Evolutionary tendencies, ii) Genetic influences, iii) Physiological factors.

A. Evolutionary Tendencies

Means inherent characters like imperfect flowers and variation in development period of the plants responsible for unfruitfulness.

a) Imperfect Flowers

Monoecious and Dioecious

- **i) Monoecious**: Flowers are monoecious (separate male and female flowers but present on the same plant) *e.g.* coconut, arecanut, banana, litchi, aonla, pecan nut and capri fig etc. Some times in the presence of pollinator there is no or very little chance of pollination and fruit setting. In case of figs two types of flower clusters are borne namely staminate and pistillate flowers. In Capri fig staminate flowers are borne near the eye while the pistillate flowers are borne near the end.
- **ii) Dioecious**: Male and female flowers bear on different plants– in these unisexual flowers, self-pollination is eliminated *e.g.* papaya, date palm, kiwi fruit, pistachio and strawberry. Likewise, a few varieties of plum produce too little pollen to call them bisexual. Profuse flowering without fruit set in ornamental pomegranate is a result of their being unisexual.

Structural peculiarities: May be having perfect flowers but the period of pollen shed do not coincide with the receptivity of pistil. This condition of maturity of stamens and pistil at different time is called as dichogamy.

Dichogamy: It is the condition of maturation of stamens and pistils at different time in bisexual/hermaphrodite flowers. In this condition the receptivity of the stigma does not coincide with the availability/viability of pollen is called dichogamy. In dichogamy, self-pollination is prevented because the maturity of both organs do not coincide.

Protoandrous: If the stamens ripe before the stigma becomes receptive the flowers are known as protoandrous *e.g.* coconut, sapota, walnuts.

Protogynous: If stigmas become receptive before the stamens produce viable pollens it is known as protogynous *e.g.* plum, banana, pomegranate, fig.

Hetrostyly: In which style and filaments are of different length lead to poor fruit set. A condition in which length of the style differ in the flowers of same or different plants in relation to other parts of the flower, it also prevents self-pollination. Means difference in length of filaments and style are not favourable for self-pollination, if length of a style is shorter than length of filament or vice-versa.

Pin type: Sapota, litchi, pomegranate.

Thrum type: Almond, litchi.

In case of delicious group of apple upright stamens position accompanied by spread of petals, which do not allow bees to pollinate the flower while collecting nectar. When the pistils of heterostyled plants pollinated with pollen of same or other flowers of equal height stamens may be fruitful or may have varying degree of sterility.

In brinjal styles are available in four different lengths *i.e.* long, medium, pseudo short and true short. Among these, pseudo short and true short do not produce fruits.

b) Abortiveness Leading to Impotence (Abortive flowers or aborted pistils or ovules)

Abortion of flowers and flower bud before blossoming is well known. Flower abortion is more common in indeterminate inflorescence as compared to determinate. In case of strawberry abortion of pistil is late, so unfruitfulness does not result. Late flowers of strawberry are abortive, so fruit can be formed by earlier flowering. Falling of terminal portion of pistillate flower before pollination in pecan, grapes and tomato lead to unfruitfulness. There are cultivars of defective pistil (plum), defective embryo (apple), defective embryo sac (oranges) leads to unfruitfulness. Abortive embryo is a common problem in apple varieties; some olive varieties have 10-60 per cent abortive embryos. Abortion of embryo sac may cause seedlessness and fruitfulness in certain instances. The interference during development and functioning may cause unfruitfulness or sterility in some varieties of grape and tomato.

c) Impotence of Pollen

When pollen is not viable the production of fruit is not possible. Cultivars of American grapes may be classed as self-fertile, self-sterile and partially self-sterile when dependent on their own pollen grains for fertilization. Sterile pollen cause degeneration processes in the generative nucleus or arrested development prior to mitosis in the microspore nucleus in the grape, peach cv. J.H. Hale, Washington Navel orange and 'Tahiti' lime, 65 per cent pollens of mango are defective.

B. Genetic Influences

Hybridity and incompatibility are two types of sterility which are directly due to genetic factors. Self-sterility is an inheritance characteristic of a plant it can develop in favourable environment.

a) Sterility and Unfruitfulness Due to Hybridity

Hybridity is associated with sterility as well as unfruitfulness. It is also responsible for seedlessness in some cultivars. Generally wider the crossing encountered greater is the degree of sterility.

i. The cross between peach and plum bears flowers in abundance but without pistils with malformed stamens.
ii. A hybrid (Pyronia) between pear and quince is seedless.
iii. Cross between sweet orange and *Citrus trifoliata* (citranges) produce infertile female gametes.
iv. Most of the banana and pineapple varieties have seedlessness due to hybrid nature of their ancestors.
v. Most of the triploid apple produce aborted pollen.
vi. The hybrids between *Vitis rotundifolia* and *Euvitis* are completely sterile.
vii. Similarly, hybridization of *Vitis vinifera* and *Vitis rotundifolia* are also completely sterile.

b) Incompatibility

Several cultivars of fruit trees though being self-unfruitful give good performance when cross fertilization take place. So, there is nothing wrong with pistil. Likewise, capacity of pollen to be effective on another cultivar of same species tends to suggest that both pistil and pollen are normal and capable of functioning except when incompatible *e.g.* fig, Apple, pear, plum, cherry, citrus, strawberry.

i) Most common reason of self-sterility or self-unfruitful is the incompatibility between the pollens and ovules of the same flowers, other flower of the same plant or other plant of the same variety, if fertile, it fails to affect conjugation.
ii) In apple, pear, plum and aonla pollinizer varieties are required to avoid self-incompatibility.
iii) Self-incompatibility in some varieties like: Don't bear if planted as sole cultivar.

   Mango: Langra, Dashehari and Chausa.

   Loquat: Late Yellow, Golden Yellow, Tanaka, California Advance.

   Lemon: Pant Lemon, Kagzi Kalan.
iv) Self-sterility and self-unfruitfulness due to incompability has been reported in many crops like apple, pears, plums, almond, apricot where normal processes of fertilization fail.

C. Physiological Influences

a) Slow Pollen Tube Growth

Sometimes due to low temperature and hormonal control (apple, pear, cherry, citrus) the growth of pollen tube remains slow in the style.

i) Slow growth of the pollen tube results unfruitfulness.

ii) Differences in pollen tube growth rate in self- and cross-pollinated apples, pears, cherries and certain citrus fruits.

iii) Chemotropic or hormone influences also cause one type of incompatibility.

iv) Fertilization process takes place within a short period of time if fail, abscission at base of the style, ovary pedicel or peduncle will take place and fruit setting does not take place.

b) Premature or Delayed Pollination

Difference in stage of maturity and fertilization leading to development of a fruit with viable seeds is controlled by the nutritive conditions within the plants existing at the time of pre-blossoming and post blossoming stage. Deficiency of carbohydrates cause failure to blossom, abortion and fruit drop as compared to spur borne fruits. Ringing leads to parthenocarpy due to increased concentration of nutrients at the point of bearing. Over bearing, over pruning, ill nutrition or unfavourable water relations lead to defective pistil or pollen grains of low viability.

i) Premature pollination in tobacco flowers is very susceptible to injury.

ii) When mature pollen grains are applied on immature pistils they germinate, penetrate into the style, enter into the ovule and if the ovules are not ready for fertilization the flowers fall.

iii) In case of oranges premature pollination did not have any deleterious effect, however, some injury was noticed in tomato.

iv) Premature pollination noticed lower fruit setting in persimmon, pear, plum and peach.

v) Delayed pollination resulted in falling of flowers without fruit setting. If pollination delayed for 1 or 2 days did not affect fruit set, but further delay may result into polyembryonic seeds in some species.

Nutritive Condition of Plant

- ☆ Nutrient status of the plant just before and after flowering is an important factor to determine the fruit setting percentage. It also affects the viability of pollen or fertility of pistil.
- ☆ Deficiency of carbohydrates at flowering is common cause of fruit drop in *Vitis vinifera*. In green house tomato crop the deficiency of carbohydrate cause flower abortion and ultimately unfruitfulness.

Effect on Pollen Viability

When pollen is collected from very old and young plant of apple, significant difference in germination of pollen was observed.

Effect on Defectiveness of Pistils

Drought conditions, less fertile soils and overbearing tree produce defective pistils. Over bearing exhaust/weaken the tree for production in ensuing season. In American plum a close correlation between defective pollen and unfruitfulness was observed.

Flowering and Fruit Setting at different Positions

- In case of terminal bearing crops, fruit setting percentage is very less due to more competition in fruit setting.
- The competition at the position takes place between fruits and branch on fruit influencing factors.

Strong and Weak Spurs

- There is a strong correlation between fruit set and the vigour of the limb, more vigorous limb bear more fruit, less vigorous or weak limb bear less fruits. Strong limbs bear more flower and fruit setting/yield as compared to weak limb. Similarly, if flowers born singly, set fruit and mature however, if flower borne in cluster majority of them will drop.
- Girdling and ringing of twig accumulate more food material to set fruit and also develop parthenocarpy.
- If nutritive conditions of the fruits are favourable the embryo accomplishes development into seed coat and fruit wall, if not favourable it retards development.
- Under nutrient deficient conditions or insufficient nutrient supply the number of seeds forming ovules diminished and under extreme nutrient deficient conditions large number of ovules are unable to form seed.
- Cucumber crop in greenhouse conditions arrest the growth of fruits under nutrient deficient conditions and position of the fruits. If some fruits are harvested or thinned out, the remaining fruit resume their full growth.
- In case of strawberries under nutritional deficiency, it produces bisexual flowers which may lead to produce pistillate flowers.

2. External Factors Associated with Unfruitfulness

Crop production is an adjustment between heredity and environment.

a) **Nutrition**: Nutritive condition is affected by soils, water supply, manuring, soil culture and pruning exert a profound influence on fruit set. Application of manures and fertilizers few days before bloom emergence is generally beneficial for proper fruit set. Excess as well deficiency of nutrients leads to unfruitfulness.

b) **Pruning and grafting**: Pruning affects flower formation, fruit setting and productivity. Moderate to heavy pruning in apple had better fruit set few days before blossom emergence. Moderate pruning induces flower formation, good fruit setting and yield in soils heaving ample available nutrients. Pruning is also known to give rise to normal true hermaphrodite flowers in grapes where absence of pruning leads to sterility due to weakening and abortion of its pistils.

Pruning and grafting affect nutrient supply through manipulation of C/N ratio.

Pruning and grafting remove apical dominance.

c) **Locality**: A cultivar may result in good fruit setting in one locality and no fruit setting in another locality due to microclimate, soil type, water quality. If the climate of a particular locality is not suitable, the fruit setting is rather poor.

d) **Environment**: Rainfall, temperature, humidity, light, wind and frost are major influencing factors these may favour or prevent the optimum pollination and fertilization.

   i) **Temperature**: Requirement of temperature for each crop flowering, fruit setting and fruit development specific and quality of fruit is also affected. Temperature also affect pollination, fertilization (slow pollen tube growth at low temperature), fruit setting. Temperature at 50°F allow free germination of pollen in plum, cherry, apple, peach but at 40°F or lower completely check the same. Adverse temperature inhibits pollinators. Excessive high temperature in arid areas causes unfruitfulness in sapota owing to drying of stigmatic fluid.

   ii) **Rain and wind**: Affect fruit setting by disturbing the process of pollination, germination of pollen and fertilization. Winds help in wind pollination but check insect pollination, mechanical injury to plants, dry stigma fluid premature. Rainfall and winds affect the bee activity.

   iii) **Humidity**: Dry air prevent germination of pollen grain. More humid increase infestation of pest and disease.

   iv) **Light**: Important for photosynthesis, flowering (short day, long day) and insect movement.

e) **Age and vigour of plant**: C:N ratio and nutritional status increased with maturity. More vigorous tree having more vegetative growth and less fruiting.

   v) **Distributed water relation**: Fluctuating soil temperature and inadequate supply of water has great impact on fruit setting. Moisture stress should be avoided during fruit setting and later stage of fruit development. Excessive moisture should also be avoided. In general, during flowering irrigation should be avoided in orchard. Citrus is very sensitive to application of irrigation during flowering as it makes the tree unfruitful.

vi) **Season:** Sometimes season of flowering *i.e.* early or late flowering may affect the pollination. Hybrid grape 'Ideal' is self-impotent when early in season but later on it becomes self-potent.

vii) **Pest and diseases**: Severe attack of hopper, caterpillar, butterfly, thrips, aphids, mealy bug affect fruitfulness badly. Many insect pest and diseases attack fruit plant at various stages and thus, affect fruit setting adversely *e.g.* mango hopper and mealy bug suck sap from stalks of developing mango fruits which later on drop off from the panicle. Likewise, powdery mildew in mango and grapes has distinct influence in fruit setting. Citrus psylla - just before flowering result in poor fruit setting in orange.

viii) Spraying at the time of flowering inhibit/reduce the activity of pollinator and fruit set. Some fungicides like copper fungicides @ 200 to 10000 ppm gave inhibitory effect on pollen grains and their germination on the stigma.

ix) **Pollinizers**: Some self-sterile and self-incompatible cultivar requires pollinizers *e.g.* almond varieties, J.H. Hale cultivar of peach. Pollinizer is recommended for pollination as third tree in third row. In deciduous fruit trees, grafting limb of each tree with pollinizers can be done.

x) **Sex distribution**: Dioecious (date, papaya, kiwi, pistachio) and staminate plants should be planted in certain ratio. Some monoecious plants like coconut, walnut, peanut, aonla are strictly cross pollinated which require pollinizer.

Steps to Overcome the Problems of Unfruitfulness

There are so many reasons for unfruitfulness in the fruit crops, it is necessary to understand the reasons and take some corrective measures from planning or establishment of orchard.

i. Selection of crop and variety should be on the basis of climatic and edaphic conditions of the orchard location.

ii. Windbreak and shelter belts should be planted before or at the time of planting to avoid wind damage.

iii. Before planting orchard soil should be incorporated with organic matter, nutrients upto a optimum level on the basis of soil nutrient status.

iv. A mixture of variety or pollinizer varieties and varieties which attract pollinator should be planted in between the orchard to avoid the problem heterostyly, dichogamy incompatibility, sterility, embryo abortion, hybridity, *etc.*

v. Standardize the use of growth regulator at appropriate time in required quantity for proper growth of pollen tube for proper pollination.

vi. Problem of old age can be overcome by new plantation or rejuvenation of old orchard.

vii. Over or excess bearing or uneven bearing can be managed by thinning of flowers or crop regulation at appropriate stage.

viii. Maintenance of proper moisture in the orchard is also helpful for fruiting as water logged or stress of moisture are harmful for flowering and fruiting.

ix. Maintain optimum level of nutrient on the basis of leaf nutrient analysis and soil testing.

x. In crops which bear flowers on current season growth require regular pruning, it should be adopted on the basis of variety and its phenology.

xi. Insect pest and diseases should be managed effectively following integrated approach. Pesticide should not be applied during the visit of pollinators.

xii. Regular bearing varieties should be planted to overcome the irregular bearing/biennial bearing.

The reason of unfruitfulness should be analyzed properly to suggest the corrective measures and to incorporate in the planning of new orchard. Package of practices are to be followed to avoid unfruitfulness.

Table 11.1: Causes of Unfruitfulness

Internal Factors			*External factors*
Evolutionary factors	**Genetic influence**	**Physiological influence**	i. Nutrient supply ii. Pruning and grafting iii. Season/locality iv. Age and vigour of plant v. Temperature vi. Light vii. Distributed water relation viii. Rain at bloom ix. Pest and diseases x. Wind xi. Spraying
i. Imperfect flower ii. Structural peculiarities Dichogamy – Protogynous - Protoanderous iii. Abortiveness leading to impotence iv. Impotency to pollen	i. Sterility and unfruitfulness due to hybridity ii. Incompatibility	i. Slow growth of pollen tube ii. Premature or delayed pollination iii. Nutritive condition with in the plant	

Overcome unfruitfulness
i. Selection of regular bearing variety
ii. Use wind break or shelter break
iii. Proper nutrition
iv. Maintain proper soil moisture
v. Grow pollinizer variety
vi. Use growth regulators
vii. Timely pruning
viii. Rejuvenate old orchard
ix. Insect pest and diseases management

Chapter 12

Pollination

Transfer of pollen from anther (stamen/male part) to stigma (pistil/female part) of a flower is called pollination. Cross pollination is essential for a number of agricultural crops and horticultural crops. The pollen can be transferred through various agents such as wind, splashing rain, honeybees, wasp bee, solitary bee, butterflies, moths, birds, depending on the plant species. More than 44 species from 15 families and 5 orders visit apple flowers. Honey bee considered the most effective pollinators. Four species of honey bee are: *Apis florae, A. dorsata, A. cerana* and *A. mellifera*. First three are indigenous and *A. mellifera* introduced in nineteenth century. *A. cerana* is the most abundant insect species visiting apple. The agents involved in pollination are called pollinators, the source of pollen is called as pollinizer. Most of the fruit crops are being pollinated by insects specially bees while nut plants pollinated by wind.

Honey bees are most effective pollinators because:

1. Its body parts are specially modified to pick up pollen grains in large number.
2. The bees show flower fidelity and consistency.
3. The bees have potential for the long working hours.
4. Maintain high population when and where required.
5. Adopt itself to different climates.

Hence, bee keeping increased fruit yield from 5 to 25 percent in addition to increasing farmers income from production of honey.

Pollinate: Transfer of pollen from one flower to another.

Pollinator: The agents (biotic or abiotic) help to transfer pollen from one flower to another.

Pollinize: Provide pollen for pollination.

Pollinizer: Plant provides pollen to fertilize.

Self-pollination: A plant that is capable of pollinating its own flowers or other flower on the same tree or flower of other tree of the same variety.

Types of Pollination

A) Self Pollination

Cleistogamy: Condition in which flower do not open *e.g.* grapes, sapota *etc.*

Homogamy: Condition in which male and female parts of flower mature at same time *e.g.* apricot, citrus, dwarf coconut, peach, phalsa *etc.*

B) Cross Pollination

Monoecious: Borne both male and female flowers on same plant but at different location not as perfect flower *e.g. Vitis rotundifolia* (muscadine grape).

Dioecious: Male and female flowers borne on separate plant *e.g.* papaya, date palm, Kiwi fruit, grape (muscadine), nutmeg *etc.*

Gynodioecious: Pistillate (female) flowers borne on same plant but bisexual and staminate (male flowers) borne on other plant *e.g.* fig.

Conditions Necessitating for Cross Pollination

Dioecious: Male and female flowers are born on different plants *e.g.* papaya, date palm, kiwi fruit.

Dichogamy: Anther and stigma of a flower mature at different time.

Protogyny: Gynoecium mature earlier to androecium *e.g.* sapota.

Protoandry: Androecium mature earlier to gynoecium.

Heterostyly: Height of the stamen and style is different.

Pin type: Flowers having long style and short filaments *e.g.* sapota, litchi, pomegranate.

Thrum type: Flowers having short style and long filaments *e.g.* almond, litchi.

Self-sterility: Pollens of the same flower unable to fuse with the stigma.

Incompatibility: Pollen is viable but unable to germinate due to genetic factor.

Environmental factors: Sometimes environmental conditions not favour the self-pollination.

Most Common Pollinators

Entomophily: Pollination by insect *e.g.* citrus, guava, ber, mango, litchi, apple *etc.*

Anemophily: Pollination by wind *e.g.* coconut, date palm, papaya, sapota *etc.*

Hydrophily: Pollination by water.

Ornithophily: Pollination by birds *e.g.* banana, pineapple.

Chiropterophily: Pollination by bats.

Anthophily: Pollination by human being.

Zoophily: Pollination by birds or insects.

Terminology

Heterogamy: Some short of barrier between stamen and pistil of the same flower so fusion of unlike gametes takes place.

Cross compatible: Two varieties are compatible for pollination and fertilization with each other.

Cross incompatible: When one variety is not receptive to pollen of the other variety.

Cross pollination: Transfer of pollen to stigma of other variety.

Fertilization: Fusion of male and female gametes to form or develop a seed.

Allogamy (Cross pollination): Pollen of one flower unites with the pollen of other flower, when variety is self-unfruitful.

Autogamy (self-pollination): Pollen and stigma of the same flower fused together.

Self-fruitful: Pollen of the same or other tree of the same variety fused together to form a seed *e.g.* peach, apricots and nactarine.

Self-unfruitful: Pollen of the same tree or the different tree of the same variety unable to fertilize *e.g.* varieties of plum De Montfort and Green Gage.

Triploid: Having three sets of chromosomes. Triploid produce good fruits but cannot serve as a source of pollen for other varieties.

Features of good pollen source variety:

- Has viable pollen that germinate well.
- Cross compatible with main variety.
- Bloom period overlap with main variety.
- Bloom at young age and bloom regularly/annually.
- Not susceptible to disease and winter hardy.
- Produce pollen at relatively low temperature/high temperature.
- Bear good, attractive and marketable fruit.

Table 12.1: Pollinizer Cultivars

Sr.No.	*Fruit Crops*	*Pollinizer cultivars*
1	Aonla	Francis, Chakaiya
2	Apple	Tydeman's Early, Red Gold, Golden Delicious, McIntosh, Winter Banana, Golden Spur
3	Plum	Beauty, Santa Rosa, Mariposa

After the process of pollination, the pollen grain germinates and grow into the flower ovary where it (male pollen) unites with female egg cell is called fertilization. The success in the process of pollination and fertilization depends upon environmental conditions as well as compatibility of pollen grain with pistil of the flower *e.g.* pollen of peach unable to pollinate apple flowers.

Fertilization: Fusion of male gametes with female gametes is called fertilization. In the fertilization process pollen grain germinate on stigma and by putting pollen tube which grow through the style into the ovule where it discharges two sperms into embryo sac. One sperm fuse with egg to form zygote and other fuse with two polar nuclei to form endosperm. The distance of pollen tube travel to reach the egg depends on the length of style, it varies from species to species.

Normally one pollen tube develops from pollen grain but may develop more also, it varies with the species.

First requirement of pollen grain is the hydration from few second to few minutes to activate the cytoplasm of pollen grains, which lead to rupture of exine. After that entire contents of pollen grain move into the pollen tube.

Important Factors for Pollen Germination and Growth of Pollen Tube

Carbohydrates

In nutrient medium sugar is necessary for germination of pollen grain as it control the osmotic pressure and serve as respiratory substrate. Sucrose is most effective sugar present in pollen.

Boron

Boron, boric acid or borate are important substrate to reduce bursting of pollen tubes as well as enhance percentage germination and pollen tube growth.

Role of Boron

Effect of boron on water relation which prevent bursting of pollen tube.

Translocation of sugar.

Influence enzyme for biosynthesis of carbohydrate.

Calcium

Ca^{++} plays an important role in pollen germination and rate and direction of pollen tube growth. Pollen grain contain small amount of Ca^{++} ion which diffuse out rapidly, but when large amount of pollen grains is present the diffused Ca^{++} is trapped between pollen grains, it is called population effect or crowding effect.

Flavonols

It has important role in pollen germination and pollen tube growth as flavonol deficient plants are self-sterile because their functional pollen unable to germinate and produce functional pollen tube.

Enzymes

Cellulose and pectinase are released when pollen placed in germination medium. Exogenous application of enzyme increases rate of tube elongation as it increases the plasticity of tip region.

Plant Hormones

Auxin and gibberellin promote the pollen tube growth.

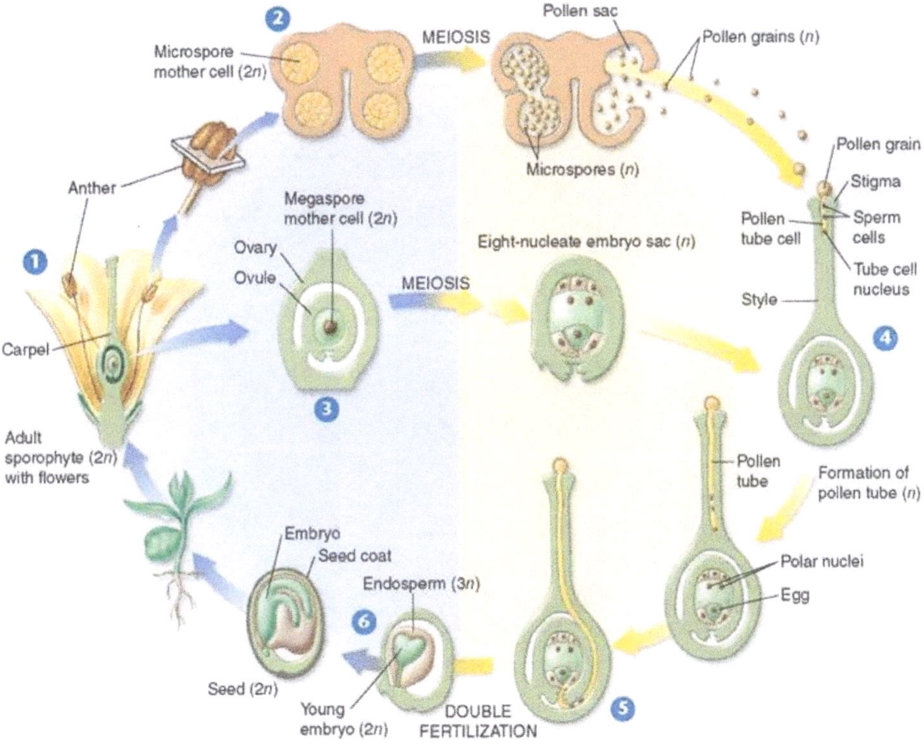

Fig 12.1: Double Fertilization (*Source: NCERT*)

Temperature

Pollen tube growth increased with temperature, the optimum temperature is 20-30°C.

Role of Stigmatic and Stylar Exudates

- ☆ Help in adhesion, hydration and germination of pollen on the stigma.
- ☆ Protect pistil from insect attack/microbial infection.
- ☆ Serve stigma for dehydration.
- ☆ Serve as food for pollinator.
- ☆ Serve as nutrients for growing pollen tube through stigma and style.

Entry of Pollen Tube into Ovule (on the basis of place of entry)

- ☆ Porogamy: Pollen tube enter through micropyle (maximum probability).
- ☆ Chalazogamy: Enter at chalazal end.
- ☆ Mesogamy: Enter through funiculus or integuments.

Chapter 13

Parthenocarpy

Parthenocarpy means virgin fruit. Parthenocarpy is the development of fruit without pollination and fertilization. The production of seedless fruit may be natural or artificial without fertilization of ovule. It is termed as formation of seedless fruits in the presence of pollination or any other stimulus. The term was introduced by Null in 1902.

The seeds from parthenocarpic fruits are not fully developed and they do not germinate.

Types of Parthenocarpy

Parthenocarpy is broadly of two types:

1. Natural or genetic
2. Artificial or induced

1. Natural or Genetic Parthenocarpy

It has been documented in various crops like grape, tomato, mandarin, banana *etc.* It is further classified into following classes.

a. Obligatory or Automatic Parthenocarpy

Parthenocarpy due to prevention of pollination. It always results in seedless fruits *e.g.* banana, pineapple.

b. Facultative Parthenocarpy

It occurs due to conditions adverse to pollination or fertilization or results from genetic sterility due to continuous vegetative propagation *e.g.* grapes, tomato, banana, pineapple *etc.*

c. Vegetative or Autonomic Parthenocarpy

Pollination or other external stimulation is not required (fruits develop without pollination) to produce parthenocarpic fruits and fertilization *e.g.* banana, apple, fig, pineapple, Washington Navel orange, Oriental persimmon, cucumber *etc.*

d. Stimulative or Aitionomic Parthenocarpy

A type of parthenocarpy in which fruit development involves pollination (pollen stimulus) but no fertilization *e.g.* banana because it is triploid (it is result of a diploid and a tetraploid parent) and therefore cannot produce seeds *e.g.* litchi, Black Corinth grapes, blackberry, pear *etc.*

e. Stenospermocarpy

It was coined by Stout in 1936 pollination and fertilization occur but the embryo gets aborted *i.e.* embryo abortion is the cause of seedlessness *e.g.* grapes, Sindhu variety of mango.

2. Artificial or Induced

Induced parthenocarpy is the production of seedless fruits by treatment of the flower with materials such as dead pollen (mentor), pollen extract, chemicals and growth substances. Plant hormones like auxins, gibberellins and cytokinins when sprayed on flowers always stimulate the development of parthenocarpic fruits *e.g.* frost and low temperature also result in production of parthenocarpic fruits *e.g.* apple and pear.

Insect attack produces small seedless fruits in some apple varieties.

Parthenocarpic fruits have been produced by bark ringing in gooseberry, grapes and apples.

Cutting of style from base before the pollen tube reaches the ovary induces seedlessness in several citrus fruits.

Chemical can be used for providing the stimulates for the development of mature fruits. Auxins like indole acetic acid, indole proponic acid, indole butyric acid and phenoxy acetic acid have been used to induce parthenocarpy in fruit crops.

The induction of parthenocarpy results in changes in physico-chemical characteristics of fruits, examples are as under:

1 **Size:** The size of seedless fruits remains smaller than the seeded fruits, however the quality remains better and mature earlier. *e.g.* size of seedless mango fruit and seedless grapes berries are smaller than seeded fruits and berries, respectively.
2. **Form:** The GA induced parthenocarpic fruits are elongated as compared to normal seeded fruits.
3. **Composition and quality:** Seedless fruits are sweeter than seeded fruits *e.g.* most of the seedless grape fruits are sweeter than seeded fruits of the same variety.

4. **Maturity:** The seeds in seeded fruits produced ethylene on maturity so these fruits ripen earlier than seededless fruits.

Importance/Advantage of Parthenocarpy

1. Fruits without seeds is a desirable trait in some edible fruits like banana, pineapple, orange and grapefruit.
2. It is desirable in crops in which pollination and fertilization is difficult *e.g.* tomato and summer squash.
3. Parthenocarpy increase fruit production or helpful in fruit setting of dioecious fruit species *e.g.* persimmon, because male/staminate plant is not required to pollinate or to provide pollen.
4. Parthenocarpy also increased the novelty of fruits like cherry, grapes, strawberry – they are easier to eat with less waste.
5. Improve taste and palatability.
6. Improve processing quality and ultimately increases the profitability in many fruit crops.
7. Seeds in the fruit produce ethylene so seedless fruits have more shelf life.
8. Increase labour efficiency in food processing industry.
9. Fruit set is not affected by the factors (environmental) which inhibit the pollination and fertilization.
10. Fruit set and development start before anthesis in parthenocarpic plants so it is helpful for early season fruit production.
11. Parthenocarpy is helpful to improve the quality and productivity of fruit crops.

Undesirable Effects

1. Parthenocarpy is undesirable for the crops in which seeds are edible *e.g.* almond, pistachio, walnut *etc.*
2. Parthenocarpy occasionally occurs as a mutation in nature and is considered as a defect because the plant can no longer sexually reproduce.

Origin of Parthenocarpy

1. Fruit set without pollination or fertilization.
2. Sterility in male flowers.
3. Degradation of pollen mother cell.
4. Degradation of fertilized ovules.
5. Irregularities in chromosomal due to triploidy (seedless) *e.g.* Tahiti lime, Oroblanco.
6. Due to self-incompatibility *e.g.* citrus.

7. Due to embryo abortion.

Causes and Induction of Parthenocarpy in Fruit Crops

1. Guava *e.g.* Allahabad seedless.

   Due to triploid.

   Failure of fertilization of well differentiated ovules.

   Degeneration of fertilized ovules.
2. Grapes *e.g.* Thompson Seedless, Flame Seedless, Crimson Seedless.
   Due to embryo abortion (stenospermocarpy).
3. Embryo rescue technique is used to produce seedless grapes. In this technique the tiny embryo (immature) is removed and grown artificially through tissue culture.
4. Citrus: Due to lack of pollination.

   Self-incompatibility (mandarin).

   Cytoplasm male sterility (Satsuma).

   Chromosomal irregularities (Tahiti Lime, Oroblanco).
5. Apple – lack of pollination.

Chapter 14

Plant Growth Regulators

Plant growth regulators (PGRs): Plant growth regulators are natural or synthetic compounds other than nutrients used in very small amount to promote or inhibit the development or metabolic processes in plants. These do not have nutritive values.

Plant hormones: Endogenous growth substances (natural).

Growth regulators: Synthetic growth substances.

Phytohormones: These hormones are used at very low concentration to regulate the physiological process of plant produced. These usually move at site of action from a site of production/synthesis in plants.

Growth retardant: The term growth retardant or growth retarding chemical are those chemicals which slows down the process of cell division and cell elongation used to regulate plant growth (height) without developmental effects *e.g.* CCC, AMO 1618, Chlormequat, Phosphon-D and Alar. These are not occurring in plant system naturally.

Growth: Quantitative and irreversible increase in size of plant such as plant height, root length and number of leaves *etc.*

Development: Qualitative changes in different phases of plants such as initiation and formation of leaves, initiation of flowers and development of fruits, shedding of leaves and fruits.

Plant hormones: These are organic compound which respond physiologically at very low concentration and synthesized in one part of the plant and translocated to another part.

Promoters

Auxins

Gibberellin

Cytokinins

Inhibitors

Abscisic acid

Ethylene

Hypothetical growth substances

Florigen

Death hormone

a) Growth Promoter

Auxin

It is derived from Greek word Auxin means to increase or stimulate cell elongation and discovered by Charles Darwin and Francis Darwin. It influences the growth and development response. Site of synthesis is the young tissues/young leaves. It promotes apical dominance and its movement is from shoot to root *i.e.* basipetal. Endogenous auxin is IAA. Auxin involved in phototropic and gravitropic reactions.

Most Commonly Used Auxins

- Indole-3-acetic acid (IAA)
- Indole-3-butyric acid (IBA)
- 1-naphthalene acetic acid (NAA)
- Phenyl acetic acid (PAA)
- 2,4-dichlorophenoxy acetic acid (2,4-D)
- 2,4,5-trichlorophenoxy acetic acid (2,4,5-T)
- Picloram dicamba

Functions of Auxin

- It stimulates the cell division and elongation.
- Helpful in differentiation of phloem and xylem.
- It used for initiation in cuttings *e.g.* IBA and IAA, successful air layering in sapota can be done by treating the cut portion with 1000 ppm IBA.
- It is used to inhibit apical dominance *e.g.* NAA.
- Parthenocarpic fruit: Grapes, banana and orange (IAA).
- Increase fruit size of plum and pineapple.

- Retard abscission in the plants.
- Control fruit drop in citrus: 10 ppm 2,4-D +0.5 per cent zinc sulphate + 20 ppm aureofungin or 20 ppm NAA +0.5 per cent zinc sulphate + 20 ppm aureofungin is recommended to control fruit drop in citrus.
- IBA @ 100-500 ppm, 500-1500 ppm and 2000-5000 ppm used for rooting in soft wood, semi hard wood and hard wood cuttings, respectively.
- Callus formation in *in-vitro* multiplication *e.g.* 2, 4-D, shoot multiplications (IBA and BAP).
- Delays fruit ripening, prevent abscission layer formation.
- Stimulates growth of flower parts: NAA is used for uniform flowering and fruit ripening in pineapple while 2, 4-D delay flowering.
- Girdling in litchi can be replaced be application of NAA.
- Thinning of bloom in apple and guava: NAA.
- Act as a weedicide to control broad leaf/dicotyledon weeds *e.g.* 2,4-D, 2,4,5-T.

Cytokinins

- Discovered by F. Skoog, C. Miller and co-workers during the 1950s.
- Its main function is cell division (cytokinesis).
- It was first discovered as adenine (aminopurine) derivative named kinetin (6-furfuryl- aminopurine), which was isolated as a DNA degradation product.
- First endogenous cytokinin zeatin was isolated from maize kernels.
- It promotes axillary branching.
- It is also called juvenile hormone.

Cytokinin like Growth Regulators

- Zeatin (Z)
- Zeatin riboside (ZR)
- Isopentenyladenine (iP)
- Isopentenyladenosine (iPA)
- 6-benzylaminopurine (BAP)
- Kinetin and Thidiazuron (TDZ)

Functions

- Induce cell division and organ differentiation which help in growth and development of plant.
- Help in seed germination and initiation of radicle.
- Inhibition of leaf senescence.

- ☆ Tissue culture morphogenesis.
- ☆ Adventitious shoot formation.
- ☆ Apical dominance overcoming.
- ☆ Stimulate axillary bud outgrowth.
- ☆ Inhibition elongation of shoot.
- ☆ Induction of flower and fruit development.
- ☆ Breaking dormancy.
- ☆ Delay senescence.

Gibberellin

Gibberellin was discovered by Japanese scientist, Eiichi Kurosawa in 1926 during study on *bakanae*, the "foolish seedling" disease in rice. Most common gibberellin is GA_3.

Functions

- ☆ Induce cell and shoot elongation/expansion: Spray of GA_3 increase the height of seedling.
- ☆ Release dormancy of seeds, somatic embryos, apical buds and bulbs.
- ☆ Decrease abscission by increasing photosynthetic and protein synthesis.
- ☆ GA sprays terminate rest period (replace dormancy and chill unit) of buds in peach and apple.
- ☆ GA accelerate the germination of seed in many crops.
- ☆ In pecan, GA significantly reduce/substitute the period of seed stratification.
- ☆ GA application produce parthenocarpic fruits like apple, pear, fig, grapes *etc.*
- ☆ Inhibition of adventitious root formation.
- ☆ GA_3 is used to increase berry size in grapes *e.g.* 20 and 40 ppm GA is recommended at full bloom and fruit setting stage, respectively to increase the berry size in grapes. Dipping bunches in GA_3 (75 ppm) for 10 seconds improve berry size in grapes.
- ☆ GA_3 @15 ppm at full bloom and repeat after 15 days control the fruit drop in jamum.

Ethylene

Identified by Neljubow D.N. during 1901. It is gaseous growth regulating compound. Ethylene may be active in alleviation of secondary dormancy also.

Functions

- ☆ For ripening of banana, mango and other climacteric fruits *e.g.* dip fully developed fruits of papaya in 500 ppm ethrel for 10 minutes for ripening.

- ☆ Help in chlorophyll degradation.
- ☆ Induce female flowers in cucurbit plants.
- ☆ Induce male sterility if applied at pollen development stage.
- ☆ Ethylene responsible for flowering in pineapple.
- ☆ Bulb and tuber formation.
- ☆ Induce abscission of leaves.
- ☆ Promotion of dormancy.
- ☆ In lemons, dipping in ethephon @ 1000 ppm to attain marketable yellow colour of fruits.
- ☆ Application of acetylene, calcium carbide, ethephon and NAA (10-15 ppm) in pineapple induce flowering.
- ☆ Ethrel is recommended for thinning fruits in mango.

Brassinosteroids

i) Brassinosteroids (polyhydroxy steroids) were recognized as sixth class of plant hormones.

ii) First explored by Mitchell *et al.,* 1979 nearly 50 years ago.

iii) Brassinolide was isolated in 1979 from pollen of *Brassica napus.*

iv) It promotes stem elongation.

Functions

- ☆ Cell and stem elongation which promote shoot growth
- ☆ Unrolling and bending of grasses
- ☆ Ethylene production
- ☆ Seed germination
- ☆ Xylem differentiation
- ☆ Pollen tube growth
- ☆ Leaf abscission

Jasmonates

It activate the stress protein synthesis.

These promote senescence of leaf, ripening of fruit and formation of tuber and bulb. Jasmonates play a role in development and breaking of dormancy. It is also noticed that jasmonates promote shoots and roots regeneration.

Inhibitors

The compound which inhibit or suppress the plants growth.

1. Phenolic inhibitors
2. Synthetic inhibitors

Phenolic inhibitors: *e.g.* Salicylic acid, chlorogenic acid, coumaric acid and benzoic acid.

Synthetic inhibitors: An inhibitor synthesized from the young leaves of *Betula spp.* prevent the growth of apical buds *e.g.* ABA and Dormin, CCC, Tri-Iodobenzoic acid (TIBA), SADH, Maleic hydrazide and benzothiazol were observed to promote flowering in lemon.

Abscisic Acid

It is best known as a plant hormone also known as Abscisin II and Dormin. The role of abscission is in the abscission of leaves, flowers and fruits. During winter dormancy it is produced in terminal buds to avoid shoot growth. During water stress and other stress conditions it is produced in roots.

Functions

- ✰ Help in stomata closure and water relation.
- ✰ Inhibit vivipary in developing seeds.
- ✰ To stop cell elongation.
- ✰ Induce dormancy and satisfy chill unit.
- ✰ Delay germination.
- ✰ Inhibit growth process (root growth, photosynthesis).
- ✰ Help in senescence and ripening process.
- ✰ Increase the resistance to frost damage.

Methods of Application

Ways to apply growth regulators is as under:

- ✰ Foliar application/Spray method.
- ✰ Soil application/Root feeding method.
- ✰ Powder form.
- ✰ Dip treatment (cuttings).

Uses of Plant Growth Regulators

1. **Plants propagation and multiplication** : The plants propagated by vegetative methods like stem, leaf cutting and by layering, for promotion of rooting in plant vegetative parts, IBA is most commonly used hormone for initiation of root followed by NAA. Cytokinins is also helpful for quick and profuse rooting in cuttings and layers. Auxins used for formation of roots in cuttings like guava, pomegranate, fig, rose, crotons, hibiscus *etc.* However, GA inhibit the formation of roots in cutting.
2. **Seed germination**: To overcome the natural dormancy of seeds, dip the seeds in GA, it removes dormancy and enhance the germination. Soaking

of French beans and peas seeds in GA@ 10-20ppm for 12 hours improves the yield and quality.

3. **Plant size control**: Hormones are used to enhance or check the growth of plants.

   Foliar application of cycocel (growth retardant) check the growth of leaves.

   Spraying of morphactin @ 10ppm reduce plant growth but increase the tuber size of potato.

   The growth retardants are used to check the growth of hedges in ornamental gardens which reduce the cost of trimming the hedges.

4. **Flowering regulation**: To avoid the ripening of pineapple during rainy season spraying 5-10 ppm spray of NAA before flowering. As the quality of rainy season pineapple fruits is very poor. Flowering in dahlia plants can be induced by application of GA @ 100-200 ppm. NAA @ 300 ppm is recommended to regulate the ambe bahar crop in guava.

5. **Sex expression:** In cucurbits such as bitter gourd, ridge gourd, cucumber, watermelon and pumpkins, which produce more male flowers than female flowers. Auxins and ethrel are used to increase female flowers for increasing fruit set. Most commonly used hormones to increase female flowers are NAA and ethrel. Ethrel @ 100 ppm in bottle gourd and 250 ppm in long melon recommended as foliar application to increase number of female flowers when sprayed at 2 and 4 true leaves.

6. **Increase fruit set and fruit growth**: Foliar application of NAA, TIBA, and PCPA during flowering increases the fruit set. Fruit/berry size of grape cv. Thompson seedless can be increased by dipping bunches (young fruits) in GA solution. Ethrel @ 250 ppm is recommended as foliar application to increase number of fruits and fruit yield when sprayed at 2 and 4 true leaves in cucurbit. GA @ 25 ppm is recommended as foliar application to increase number of fruits and TSS when sprayed at 2 and 4 true leaves in water melon.

7. **Control fruit drop**: Foliar application of NAA @10-20 ppm or 2,4-D @ 10 ppm after fruit set increase fruit retention/decrease fruit drop in Nagpur santra and mango.

8. **Fruit thinning:** To make the appropriate balance between the source and sink. Spraying of ethrel or morphactin reduce the fruit load 25-30 per cent. In Guava to avoid *ambe bahar* crop NAA @300 ppm is applied for flower drop.

9. **Early ripening and development of fruit colour:** Early bearing crop fetches good price in the market as compared to late bearing crop. Maturity of apple is hastened with the application of 2,4,5-T and B-9 by 1-4 weeks.

10. **Sprout prevention**: In some crops like potatoes and onions sprouting deteriorate the quality of the produce and makes them unfit for cooking. To prevents sprouting, spray maleic hydrazide (MH) solution before storing, it will be helpful for safe storage up to 6 months.
11. **Control of weeds**: Broad leaf weeds successfully controlled by spraying of 2,4-D in many crops.

Chapter 15

Irrigation and Fertilizer Application

Irrigation

Artificial application of water to the plants during shortage of natural rains in order to obtain rapid growth and increased yields. Frequency of irrigation depends on topography, soil characteristics. Weather conditions, availability of water and kind of plant (age of the plant, root distribution, water absorption capacity, growth habit/stage *etc.*). Efficiency of irrigation is the output from the plant per unit application of water, it depends on many factors such as frequency of irrigation, duration of irrigation, intensity of irrigation, source of irrigation and methods of irrigation.

- It helps to release nutrients in the soil solution for absorption/uptake by plants and leach down or wash out injurious salts from the soil or root zone.
- It creates micro-climate by maintaining temperature and humidity.
- It maintains the optimum population of soil microbes.
- Irrigation influence several factors like photosynthesis, respiration, absorption, translocation and utilization of mineral nutrients *etc.*
- Shortage and excess of irrigation directly affects the growth and development of a plant which reflect on yield and quality.

Critical Stages for Irrigation

Just after manures and fertilizers application and during fruit setting and development. Insufficient irrigation at fruit development reduces the size, yield and increase fruit cracking and fruit drop. In a newly planted orchards or young plants, the plant should be irrigated before the young leaves starts/show wilting appearance or old plants shed their old leaves.

Irrigation Systems

There are so many irrigation methods which are adopted for different crops in different parts of the country. The irrigation method helpful for better crop growth with maximum water use efficiency, minimum seepage and evaporation losses is the best method. Water use efficiency of the conventional methods is 30-50 per cent.

Irrigation systems are divided into three broad categories:

1. **Surface irrigation**
    a. Flood type
    b. Basin type
    c. Ring/modified basin type
    d. Furrow type
    e. Drip or Trickle irrigation
2. **Sub-surface irrigation**
    a. Perforated pipelines, pitcher
    b. Trench method
3. **Overhead or aerial irrigation**
    a. Sprinkler/fountain irrigation
    b. Revolving nozzles pop-up irrigation, rain gun

1. Surface Irrigation

a) Flood Type

- ☆ When plenty of water is available and land is leveled with a slight slope.
- ☆ It is the easiest method to irrigate the orchard and permit inter-culture operations.
- ☆ This method has wastage of water and encourages growth of weeds, water wastage, soil erosion.
- ☆ The spread of diseases along with flow of water like gummosis in citrus, root rot and collar rot in papaya.
- ☆ More water loss and water use efficiency is very less.
- ☆ Spread weed seeds and flow of nutrients with flow of water.

b) Basin Type

- ☆ In this system circular basin are prepared under the canopy of the tree around the trunk. These basins are interconnected through a channel.
- ☆ There are chances of spread of the disease (like gummosis, root rot, collor rot) due to flow of water from basin of one tree to other.

- There are chances of flow of nutrient applied in the basin, these may be carried out with the water flow.
- This system requires less labour to irrigate the field.

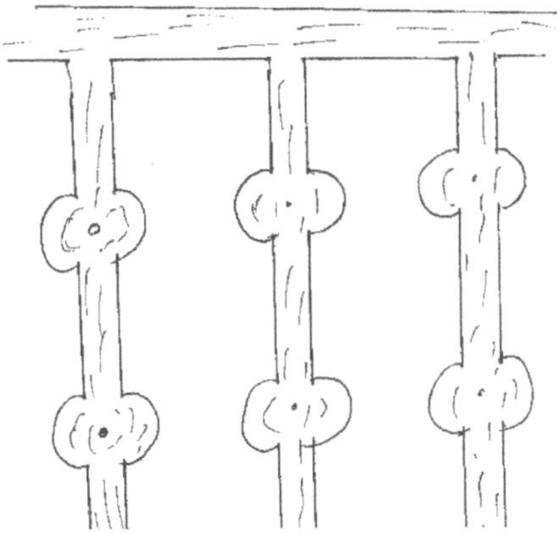

Fig 15.1: Basin System

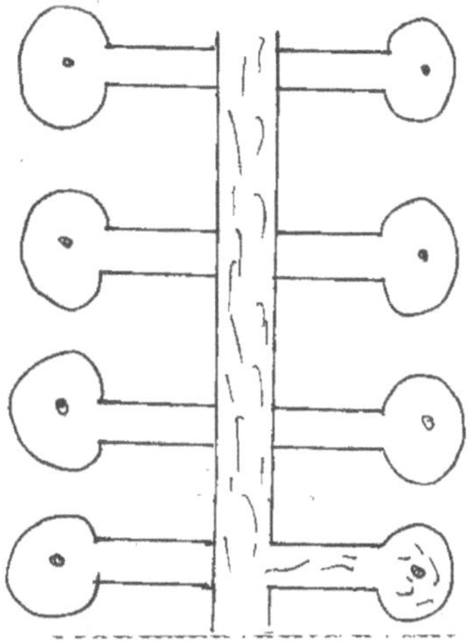

Fig 15.2: Modified Ring Basin

Fig 15.2 (a): Modified Ring Basin

c) Ring/Modified Basin Type

- It is modification over the basin system.

- ☆ In this method of irrigation, channel is prepared between the row and basins are connected independently with sub-channel. It reduces the chances to spread of soil borne disease.
- ☆ The size of the ring can be increased with the increase in the canopy size.
- ☆ This system does not permit the flow of nutrient and water from one basin of tree to other basin but require more labour and capital.

d) Furrow Type

In this method, irrigation is applied in furrows. This method is more suitable for bulbous crops. Furrows are irrigated 1/3rd to ½ of its depth.

Uneven application of water, flow of weed seed, nutrients and chances of spread of disease are more in this method.

e) Drip or Trickle Irrigation (Designed and developed by Symcha Blass in Israel, 1959)

- ☆ This save water and is most efficient method of irrigation. Water use efficiency of this method is about 90-95 per cent and labour requirement is negligible. This method also increases the yield of the crop by 21-67 per cent, it varies from crop to crop.
- ☆ In this method the drip emitters are fitted very near to the root zone of the crops. In this method water is applied on the targeted area so leaching of water/nutrients, and chance to emerge the weeds is also minimum.
- ☆ In this method water can be applied on daily basis as per the requirement of the crop.
- ☆ Low quality water can be used efficiently.
- ☆ Increase growth, yield and quality due to optimum availability of water.
- ☆ Suitable for undulated soils.
- ☆ Ease to apply fertilizer, or pesticides or other chemicals as fertigation/ chemigation.
- ☆ Fewer incidences of disease and less chances to spread disease.
- ☆ The establishment cost of this irrigation method is high so it is being used for high value crops such as fruits and vegetables, especially in crops grown in green/glass houses.
- ☆ Damage of drip line by rodents.
- ☆ Blockage of drippers by salts and algae.
- ☆ Wastage of land for preparation of channels and sub-channels is nil.

2. Sub-surface Irrigation

a) Perforated pipelines, pitcher

- ☆ In this method water is applied in furrow or underground ditches with perforated pipelines to provide sufficient water in the root zone, water use efficiency is more as compared to flood irrigation method, however, reduces root aeration.

- ☆ This method is not recommended in the soils having salts in subsurface (saline or alkaline soils).
- ☆ This method is applicable in the well leveled land otherwise it will supply uneven water.

Pitcher

The pitcher filled with water and buried in the periphery of individual tree where feeding roots are confined. The water is released slowly and slowly through the micropores available in the wall of pitcher, and take care of water requirement of the plant. The number of pitchers can be increased with the size of the canopy.

b) Trench Method

Water is applied below the soil surface through trenches and water directly reaches to the root zone. Water is moved lateral and upward through the capillary.

3. Over-head or Aerial Irrigation

- ☆ Water is applied on the canopy of the plant like rainfall at low or high pressure.
- ☆ Water can be applied through perforated pipe or nozzles.
- ☆ Loss of water in sandy soil is more due to more infiltration rate.

a. Sprinkler Irrigation

- ☆ This is the most successful method in undulated sandy soils. Water use efficiency of this method is about 70-75 per cent, however initial cost of this irrigation method is high.
- ☆ This method is efficient to irrigate the crop at seedling.
- ☆ This method creates suitable microclimate for the crop.
- ☆ This method is applicable in the low water output capacity tube well.
- ☆ Fertilizer can be applied through this method.

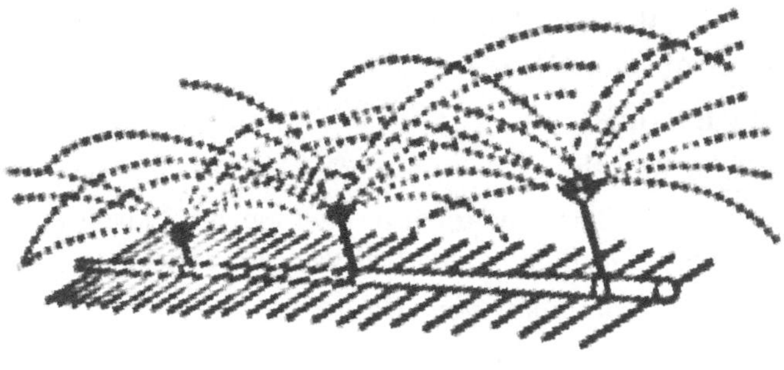

Fig 15.3: Sprinkler Irrigation Method

- ☆ It affects the pollination, lodging of crop, blockage in the nozzles and requires labour while changing or resetting the pipeline. Sometimes there are chance of blocking of pipeline due to ice formation during winter season.

b. Revolving nozzle pop-up irrigation, rain gun

This method of irrigation is used to irrigate the lawn to supply equal amount of water. The main line remains underground or fixed. The height of pop-up head is 15-20 cm and it disappears when not in use. However, in ran gun water is supplied as high-pressure sprinkler and two rain guns are sufficient to irrigate one acre.

Manuring

Application of organic waste material of plant or animal origin to add fertility of the soil. It improves the soil properties and its continuous application improves organic matter and soil fertility.

Compost

Mass developed after rotting of organic matter.

For a successful orharding, application of manure as well as chemical fertilizers at right time, in proper dose, and in right manner is essential. Fruit trees remove/ uptake large amount of nutrients from the soil. Manure requirement of fruit trees is governed by factors like age of the tree, type of soil, climatic condition, cultural practices, bearing habit *etc.* Two types of manuring:

A. Organic Manuring

FYM, poultry manure, compost, vermicompost, neem cake, oil cake, and green leaves are chief organic manures. Organic manures improve the physical conditions of soil in addition to supply of nutrients.

FYM: N = 1.5 per cent, P = 0.9 per cent, K = 1.3 per cent.

Poultry manure: N=1.5-1.8 per cent, P=1.4-1.8 per cent, K=0.8-1.8 per cent.

Green manuring: Ploughing of a cover crop *e.g. dhaincha.*

Biofertilizers: Biofertilizers contain live or latent cells of efficient strains of nitrogen fixing, phosphorus solublising or cellulolytic microorganisms used for application to seed or composting area to increase the number of such organisms.

a) **Nitrogen fixing:**

   Symbiotic: Rhizobium.

   Non-symbiotic Bacteria: **Aerobic**: *Azotobacter, Azospirillum*; **Anaerobic**: *Clostridium, Chlorobium*; **Facultative anaerobic**: *Bacillus.*

   Blue green algae: *Anabaena, Nostoc.*

b) Phosphate solubilizing microorganism/bacteria (PSM or PSB) groups of fungi and bacteria.

c) Sulphate dissolving bacteria.

d) Integrated biofertilizer.

e) Plant growth promoting rhizobacteria: *Pseudomonas, Azotobacter, Bacillus.*

B. Fertilizers with Nutrient Composition (Inorganic)

Urea ($CO(NH_2)_2$) : 46 per cent N (organic fertilizer)

DAP: 18 per cent N and 46-48 per cent P_2O_5

SSP: 16 per cent P_2O_5, DSP: 23 per cent P_2O_5, TSP: 46 per cent P_2O_5

MOP: 60 per cent K_2O

SOP: 48 per cent K_2O

- ☆ Nutrients are the chemical elements which are absorbed by the plants in more or less quantity to transform light energy into chemical energy and to keep up plant metabolism for synthesis of organic materials.
- ☆ These are synthetic or natural materials applied to soil or plant as a source of nutrient.
- ☆ Feeding of plants with nutrients is termed as nutrition.

For successful growth and production, a plant requires 16/18 elements. These elements are considered essential for life of higher plants.

Arnon (1954) laid out the following criteria for essentiality of nutrients:

1. The plant cannot complete its life cycle without these.
2. Deficiency of essential element cannot be compensated by another element.
3. The element play direct role in the plant metabolic symptom.
   - ☆ Besides these some nutrients like Aluminium (Al), Cobalt (Co), Sodium (Na), Silica (Si) and Vanadium (V) are not considered necessary always because either their essential character has been proved only in some plants or in certain metabolic processes that are not always necessary.

   1. **Basic elements (3):** Carbon (C), Hydrogen (H) and Oxygen (O).
   2. **Macro elements (6): i) Primary (3):** Nitrogen (N), Phosphorus (P) and Potash (K), **ii) Secondary:** Calcium (Ca), Magnesium (Mg) and Sulphur (S).
   3. **Micro elements (9):** Manganese (Mn), Molybdenum (Mo), Chlorine (Cl), Zinc (Zn), Boron (B), Copper (Cu), Iron (Fe), Nickel (Ni) and Cobalt (Co).

Macro elements: The elements required in relatively large quantity are termed as macro elements.

Micro elements: The elements required in relatively less quantity are termed as micro nutrients.

Table 15.1: Classification of Nutrients on the Basis of Mobility in Plant and Soils:

Mobility in Soil			*Mobility in Plants*		
Mobile	*Less Mobile*	*Immobile*	*Mobile*	*Less Mobile*	*Immobile*
NO_3^-, SO_4^{2-}, BO_3^{2-}, Cl^-, Mn^{2+},	NH^{4+}, Ca^{2+}, Mg^{2+}, K^+, Cu^{2+}	H_2PO^{4-}, HPO_4^{2-}, Zn^{2+}	N, P, K	Zn	Ca, B

Criteria for Assessment of Nutrient Requirement in Plants

a) **Growing plants in controlled conditions:** In this method the plants are grown in randomized block design and different doses of nutrients are applied to the plants. The nutrient dose at which plant give good economic returns and perform best in respect of growth, yield and fruit quality, is recommended.

b) **Soil analysis:** On the basis of soils nutrient status the dose of the nutrients is recommended.

c) **Removal of nutrients by crop (uptake):** It also gave rough idea about the requirement of the nutrient for particular crop.

d) **Leaf analysis:** Leaf nutrient status is estimated to know the deficiency/ requirement of the nutrients of a given crop. In this method sampling method and leaf types are need to be standardized.

Table 15.2: Keys to Indentify the Nutrient Deficiency

Old leaf	Necrosis (K)		
	Chlorosis	Interveinal (Mg)	
		Uniform	Early flowering or leaf abscission (N)
			Deep green foliage or few longer roots (P)
Middle part	Chlorotic leaf margin than necrosis (Mo)		
	Uniform chlorosis or faded flowers (s)		
Young leaves	Curling and Necrosis	Small leaves and short internode (Zn)	
		Smaller lighter flowers (cu)	
	Necrotic and distorted	Distortion turning necrotic (Ca)	
		Abortion of growing tip, short inter node thick leaves (B)	
	Chlorosis	Interveinal chlorosis and turning yellow (Fe)	
		Intervenial chlorosis tan speckling (Mn)	

Types of Fertilizers

Inorganic Fertilizers

- Chemicals manufactured industrially.
- Contains higher concentration of nutrient than organic manures.
- Nutrients lost through leaching, run off, volatilization and fixation by soil or consumption by weeds *etc.*

Organic Fertilizers

- These are decomposed waste of plant and animal; and are used as nutrients.
- Improves the soil tilth, aeration, water holding capacity and activity of micro-organism.

Time of Fertilizer Application

- Maximum requirement of nutrients by the plants during vegetative growth, differentiation of flowering buds, fruit setting and fruit development.
- Organic manure; about 3 months before growth initiation.
- Nitrogen: 2 weeks before growth initiation.
- 2nd dose after fruit set.

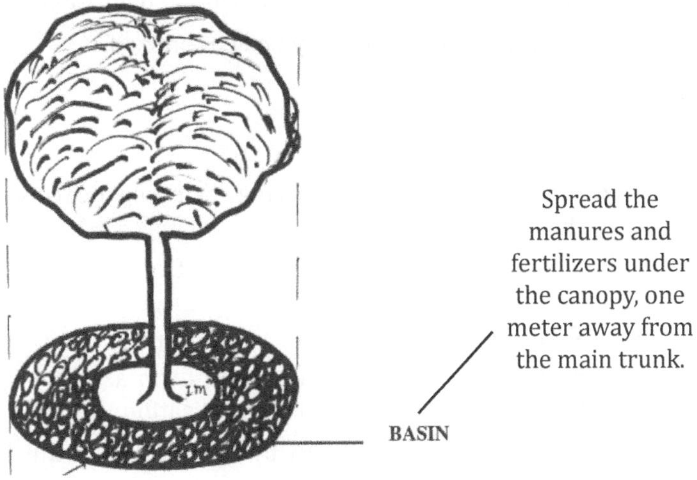

Fig 15.4: Method of Fertilizer Application

- P: 2 weeks before start of growth.
- K: In spring 4-5 weeks before fruit maturity for better quality.

- These should be applied as per the requirement of plant.
- Application timing depends on the type of fertilizer and climate.
- Nutrients should be available before or at the time of fruiting.

Methods of Fertilizer Application

It is depending upon age of the plant, root spread, type of orchard management, kind and amount of fertilizer to be applied.

i. Broadcasting

Fertilizer in solid state or granular or dust are spread uniformly over the entire field, however, leaching losses may be more.

Disadvantages

- The element like phosphorous do not readily move in the soil. Therefore, surface application may not be applicable to the trees especially in drier tracks.
- Leads to accumulation of potassium on soil surface beyond detrimental levels causing injury to plants.
- Surface application always stimulates weed growth.

ii. Band/Strip Placement

- Application of fertilizer on the sides of plant rows.
- Fertilizer in solid and liquid forms can be applied.
- Quantity of fertilizer may be economized.

iii. Ring Placement

- Commonly followed in fruit trees.
- Fertilizers are applied in a ring encircling the trunk of the trees extending the entire canopy.
- It is more labour intensive and costly.

iv. Foliar Application

- Fertilizers are applied in liquid form as foliar sprays. Water soluble fertilizers are used for foliar application.
- These are absorbed easily by leaves and give quick result.
- These are applied in a low concentration.
- Recommended when soil nutrient is not available/require early results/ nutrients are required in small quantity.

v. Starter Solution

- Application of fertilizers in liquid form.
- Roots of seedlings and propagules are dipped for varying duration in starter solution.
- The starter solution is prepared by mixing/dissolving concentrated fertilizer at a concentration not exceeding 1.0 per cent.

vi. Fertigation

- Application along with irrigation water in either open or closed systems.
- Basically, nitrogen and sulphur are applied through fertilization.
- Phosphorous fertilizers are not commonly used in fertigation because of formation of precipitates due to high concentration of Ca and Mg in water which resulted in clogging.

Advantages

- Nutrients especially (volatile or highly water soluble) nitrogen can be applied in split doses to increase its use efficiency and decreases losses.
- Nutrients utilized more efficiently when mixed with water and applied directly near the root zone.
- Cost on labour is saved.
- Best results of fertigation are noticed when the fertilizer is applied in the middle of the irrigation period *e.g.* first apply simple irrigation (without nutrient solution) for little period after that apply fertigation and just before the end of irrigation again supply simple irrigation. Use of soluble fertilizer in fertigation improves use efficiency.

Note: The grower must calculate the economics and its advantages before applying fertigation.

Hi-Tech Methods

vii. NFT (Nutrient film techniques)

Application of nutrient in the form of thin film in soil less culture media. It was developed by Dr Allen Cooper in late 1960's.

viii. Aeroponics

Spray of nutrients in the form of fine droplets on the plant and its root system. Plants are hanged in the air and rooting medium is not used to support the plant.

ix. Hydroponics

Nutrients are applied in the root zone of plants in the form of thin film or flow of nutrients.

Factors Favouring Nutrients Absorption and Transport

- ☆ High humidity, proper temperature and incident radiation.
- ☆ Good carbohydrates supply and vigorous growth.
- ☆ Chemical and physical properties of nutrient spray solution.
- ☆ Leaf characters like leaf thickness, hairyness and wax coating on the leaf.
- ☆ Generally, more vigorous plant and young growing leaves have good capacity to absorb nutrients.
- ☆ Nitrogen- applied in the form of urea (1.0 per cent) is readily absorbed.
- ☆ Sodium and potassium (KCl) - readily absorbed by leaves as these are among the highly mobile elements.

Note

- ☆ Foliar application proves to be most effective where problems of nutrient fixation in soil exits. So far, the most important use of foliar sprays is in application of micronutrients.
- ☆ Foliar sprays should be applied either with pressure sprayer or with specially designed spray guns. The trees should be sprayed until the nutrient solution begins to drip from the leaves.
- ☆ Foliar application of urea has been found effective in many fruit crops like citrus, guava, apple *etc.*
- ☆ Potassium spray (3-5 g/l)- Papaya, pineapple, citrus and guava.

Precautions

- ☆ While applying foliar sprays, care should be taken to ensure correct concentration of spray solution.
- ☆ Apply in the morning or evening hours during clear sky.

Problems/Disorder due to Nutrients Deficiency

Zn Deficiency

Guava: Rosette -Narrow and small chlorotic leaves results in die back and small sized fruits.

Apple: Rosette leaf.

Little leaf and chicken and hen syndrome in grapes.

Bronzing in guava.

Mottle leaf in citrus, coconut, litchi.

Ca Deficiency

Soft nose in apple.

Bitter pit in apple.

Cork spot in pear.

Fruit cracking in cherry.

Calyx end rot in grape and persimmon.

B Deficiency

Internal necrosis in mango.

Fruit necrosis in aonla.

Fatio in guava (red spot on young leaves).

Crown chocking in coconut.

Fruit cracking in pomegranate.

Mechanism Processed in Nutrient Transport:

1. **Passive absorption**: This type of absorption takes place without any use of metabolic energy. It may be achieved by some physical processes like diffusion, mass flow and contact exchange or root interception.

   Diffusion: Movement of nutrients from the zone of higher nutrient concentration to the zone of lower nutrient concentration.

   Mass flow: Movement of nutrient ions along with flow of water. Due to transpiration pull, nutrient along with water is absorbed by the root along the suction gradient developed between root surface and soil, suction take place without any use of metabolic energy. The greater the concentration of a solute in the soil solution, the greater the quantity brought to the root surface.

   Root interception: Roots are rapidly grown towards the surrounding areas of higher nutrient ion concentration where they intercept the nutrient ions.

2. **Active absorption**: The absorption takes place against the concentration gradient (against the ionic and chemical potentials across the membrane). It is active process because it requires the use of metabolic energy (ATP).

Index

9 789359 195520

Printed at Repro India Ltd.